Sources and Studies in the History of Mathematics and Physical Sciences

For further volumes:
http://www.springer.com/series/4142

Michael N. Fried

Edmond Halley's Reconstruction of the Lost Book of Apollonius's *Conics*

Translation and Commentary

Springer

Michael N. Fried
Program for Science and Technology Education
Ben-Gurion University of the Negev
Marcus Family Campus
Beer-Sheva 84105
Israel
mfried@bgu.ac.il

ISBN 978-1-4614-0145-2 e-ISBN 978-1-4614-0146-9
DOI 10.1007/978-1-4614-0146-9
Springer New York Dordrecht Heidelberg London

Library of Congress Control Number: 2011933227

Printed on acid-free paper

Springer Science+Business Media (www.springer.com)

For my parents Dan and Thelma

Preface

The present translation of Halley's reconstruction of Book VIII of Apollonius's *Conics* was carried out over the course of several years. It took that long because I wear another hat; namely, that of a lecturer in a program for science and technology education. And, in that field, translating and commenting on early 18th century reconstructions of ancient Greek mathematical works, unfortunately, has a somewhat low priority.

Still, though my interest in Halley's reconstruction of *Conics* VIII grew originally out of previous work I had done on Apollonius of Perga, the more I worked on the translation and thought about Halley, the more I began to see that the project actually spoke to my educational interests as well. I have long been concerned with the role that history of mathematics can play in mathematics education. I have had to ask myself in this connection what exactly does a student of mathematics gain from history? This was the question that preoccupied me more and more about Edmond Halley. Though he was a master of the modern and increasingly powerful mathematics of his time, Halley treated the ancient mathematicians with great seriousness. Why? What did he see himself learning from engaging with mathematicians such as Apollonius?

This really became for me the main question in the background of the present work. Of course one should also ask how far Halley's reconstruction of Book VIII succeeded in reproducing Apollonius's own thought and this lost book of the *Conics*. After all, that was Halley's immediate goal, and what he produced in this regard was hardly trivial. Indeed, as a person who has spent many years studying the *Conics*, I am tremendously impressed by the profundity of Halley's understanding of Apollonius. Ultimately, Halley's reconstruction of Book VIII is more about Halley than about Apollonius. It is, in a way, the portrait of this man's deep relationship with the past.

As I remarked at the outset, my own historical work has often had to take a back-seat to other institutional obligations. During a period when I felt this particularly acutely and despairingly, I was urged to take up the Halley project by a dear friend and, by all accounts, a brilliant anthropologist, Tania Forte. Tania persuaded me that without such a project I would drift inexorably, as a historian of mathematics, toward the River of Lethe. So, to Tania I must offer my very first thanks. Sadly, Tania will never read this: she died suddenly and tragically in 2005, just as she was beginning her own promising career and just as I was making my first steps in this translation and commentary. After Tania, I must thank next John Neu from the history of science library at the University of Wisconsin who generously made the text of Halley's reconstruction available to me when I was working on my Ph.D. some ten years ago. I am still touched by the kindness he showed me then. I also want to thank Gideon Freudenthal, who managed to find for me Halley's preface to the entire 1710 edition of the *Conics*, which I could not find in the best libraries here in Israel. Mayer Goldberg helped me obtain papers on Halley's actuarial work and, more importantly, provided invaluable help in all things digital, not to speak of wonderful long hours of conversation over coffee. Throughout my work on Halley's Book VIII, Sabetai Unguru was my constant advisor: he was always the first person I turned to when I had questions, doubts or ideas that needing working through and he never failed me. I owe to him deep thanks, not only for this, but also for his encouragement and warm friendship.

Thanks are also due to Marinus Taisbak who never turned me away when I came to him with Latin problems. I also want to thank my friend and fellow historian of Greek mathematics, Alain Bernard, with whom I discussed the reconstruction on several occasions. The introduction below owes much to my conversations with Alain. Lennart Berggren, who read the manuscript with a fine-toothed comb, also deserves thanks: his many comments and suggestions have made this a much better work than it was when he first received it.

Finally, I am grateful to my wife, Yifat, who truly supported me in this project, saying often, "I think you should put everything else away today and work on Halley." Although the work might have been begun without her, without her, it would never have been completed.

Beer-Sheva
April, 2011

Michael N. Fried

Contents

Part I
Introduction

Chapter 1
Edmond Halley: Ancient and Modern

Edmond Halley, whose surname was sometimes spelled *Hawley* and may well have been pronounced accordingly,[1] was born in 1656, apparently on the 29th of October. One can be confident about the year, but there is some uncertainty about the month and day. For one, records in 17th-century England did not always make clear whether a date was given by the Gregorian calendar or the older Julian calendar; also, Edmond Halley and Anne Robinson, the parents of our Edmond Halley, were married at St. Margaret's Church, it seems, only seven weeks or so before October 29th 1656,[2] making the birth date of Edmond the son, suspect, but then it might be the wedding date which is in doubt.

There is a certain irony in all this inexactness, for Halley was a man not only whose astronomical work but whose approach to understanding the world altogether was marked by careful and precise measurement. Indeed, his birth in 1656, whether or not on the 29th of October, places him squarely in the later part of the so-called "scientific revolution," when number, measurement, and mathematics assumed that central place in science they have had ever since, the period one associates with the science of Isaac Newton.

Halley probably should not be placed in the same class as Newton; yet, among the less god-like geniuses, Halley was extraordinary, and, with his talents, interests, intellectual energy and activities, he could easily stand as a representative of late 17th- and early 18th-century British science. In that regard, it is perhaps symbolic that Halley died in 1742, exactly a century after Newton was born—as if to round off the period in an appropriately precise way.

In a sense, Newton himself was Halley's greatest achievement. Halley was certainly among Newton's most important promoters: he prompted Newton to write the *Principia,* smoothed over controversies connected with it,[3] and made sure it was published. As De Morgan famously said, "But for [Halley] in all probability, the work would not have been thought of, nor when thought of written, nor when written

---

[1] It is, however, Edmond, and not Edmund.

[2] For discussions of Halley's birth date, see Ronan (1969), pp. 2–3, and Cook (1998), p. 32.

[3] Robert Hooke had claimed priority for the idea of the inverse square law.

M.N. Fried, *Edmond Halley's Reconstruction of the Lost Book of Apollonius's Conics*, Sources and Studies in the History of Mathematics and Physical Sciences,
DOI 10.1007/978-1-4614-0146-9_1, © Springer Science+Business Media, LLC 2011

printed."[4] But Halley could also boast of his own achievements. One, of course, is the comet named after him. And this was a great achievement, though the detailed observations and calculations involved in establishing the comet's periodicity[5] are not always appreciated by those for whom "Halley" *is* "Halley's comet." As an astronomer, his reputation was already established in 1678 when, working from the British territory of St. Helena and using telescopic instruments,[6] he produced at the age of 22 what was then the most complete catalogue of the southern stars. About the same time, he began to work out the details as to how observations of the transit of Venus could be used to determine the Earth's distance from the Sun. Throughout his life, and especially when he served as Astronomer Royal,[7] he made contributions to astronomy that "...touched almost every branch of the subject."[8] He improved lunar and planetary tables, described the secular acceleration of the moon and the proper motion of the stars, charted eclipses, inquired into the nature of nebulas and auroras. Outside of astronomy, he mapped terrestrial magnetic variation and in the process invented the technique of *isolines*,[9] studied trade winds and tides, developed and tested a diving bell, pioneered actuarial practices and put the calculations of this science on firm footing.[10] In mathematics one can point to Halley's procedure for calculating roots of equations (still called "Halley's method"),[11] his work on logarithms, and his theorems concerning the *loxodrome* or *rhumb line* in navigation.

With all this, one must resist thinking of Halley as a thoroughgoing modern, an ever forward-looking thinker, prophet of modern science. Moderns in Halley's time, in England and on the Continent, were, without a doubt, beginning to feel their power. For this reason, precisely, the 17th century saw a debate between supporters of ancient and modern learning, the debate immortalized in Swift's satire, *A Full and True Account of the Battle Fought Last Friday between the Ancient and the Modern Books in St. James's Library* (1704).[12] Yet Halley should not be seen

[4] De Morgan, A. (1847), "Halley," in *Cabinet Portraits of British Worthies* (ed. Charles Knight), vol.II, p. 12 (quoted in Cook (1998), p.178).

[5] Summarized in his work of 1705, *Astronomiae Cometicae Synopsis*. This was published simultaneously in English as *A Synopsis of the Astronomy of Comets*. The unbreakable association between Halley and comets was not formed because of this work, nor in his lifetime, but only after the return of the 1682 comet on Christmas day 1758, as Halley predicted.

[6] In 1678, this was still not something to be taken for granted. See below.

[7] From 1720 until his death in 1742.

[8] Berry (1961), p. 253.

[9] *A New and Correct Chart shewing the Variations of the Compass in the Western and Southern Oceans as Observed in ye Year 1700 by his Maties* [i.e., Majesty's] *Command by Edm. Halley* (1701). Halley called these "curve-lines," but they were called already in his own time "Halleyan lines" (see Warntz and Wolff (1971), pp. 79–105).

[10] See Heywood (1985).

[11] See, for example, Traub (1964).

[12] One must recall that Swift's famous satire comes at the tail of a longer debate on ancients and moderns of which some landmarks are Fontenelle's *Digression sur les anciens et les modernes* (1688), defending the moderns; William Temple's *Upon Ancient and Modern Learning* (1690) on

as a resolute captain in the modern camp—and, in this regard, he was not alone among other great scientists of his time. Katherine Hill has, for example, referred to Isaac Barrow and John Wallis as "neither ancient nor modern,"[13]and this is how one ought to refer to Halley as well, or, perhaps, refer to him as *both* ancient and modern. Halley had a deep interest and appreciation of the past.[14] When he entered Queens College, Oxford, he came, as Aubrey says, "... well versed in Latin, Greek, and Hebrew,"[15]and that was more than a mere ornament of a classical education: he valued and used his knowledge. Referring to his youthful *Commonplace Book*, Cohen and Ross (1985) say:

> Aside from the Scriptures, to which there are scores of allusions, [Halley's] principal sources are Pliny's Natural History and Aristotle's *Animalium* and *Meteorologica*. Also influential are Vitruvius, Manilius and Aulus Gellus, Strabo and Seneca, the church historian Eusebius, the church fathers Lactantius, St Jerome and St Augustine, the rhetoricians Cicero and Quintilian, the poets Homer, Horace and Virgil. Transcriptions from the ancients are characteristic of the commonplace book traditions, but the particularity of Halley's classical preferences, identified both in his youthful *Commonplace Book* and in the sale catalogue of his comprehensive library, reveals his lifelong fascination with the varieties of natural experience. (pp.1–2)

As the closing words of this quotation suggest, Halley's interest in classical sources meshed closely with his more modern scientific interests.[16] Thus his work on scriptural chronology and such studies as his 1691 work dating Caesar's landing in Britain[17] applied Halley's deep knowledge of astronomy. Conversely, much of his astronomical work—his work on the proper motions of stars, for example—arose from attentive nonperfunctory readings of historical sources. It is in this confluence of ancient and modern that one must approach Halley's work on Greek mathematical texts, particularly those of Apollonius of Perga: his translation from Arabic into Latin of Apollonius's *Cutting-off of a Ratio* and his reconstruction of *Cutting-off of an Area*, his edition of Apollonius's *Conics*, and his reconstruction of Book VIII of the *Conics*. In this connection, it is worth remarking that Halley's edition of the *Conics* with its reconstruction of Book VIII happened to be published in 1710, the very year in which Swift republished the *Battle of the Books* with a new apology for the work, showing thus that the battle was still on.

the side of the ancients; William Wotton's *Reflections upon Ancient and Modern Learning* (1694), responding to Temple and giving the edge to the moderns.

[13] Hill (1996 and 1997).

[14] According to one estimate, Halley's work on classics and other historical work constituted 13% of his professional life; by contrast, his purely mathematical work constituted only 11% and his cometary work only 3% (see Hughes (1990), p. 327).

[15] Aubrey (1669–1696, 1957), p. 121.

[16] This relationship has been discussed thoroughly in Chapman (1994).

[17] Halley (1691).

In the following pages, by way of introduction to Halley's reconstruction of Book VIII of the *Conics*, translated here from Halley's Latin text, I shall first provide some background about Apollonius's *Conics*[18] and about the circumstances surrounding Halley's reconstruction of the lost Book VIII. Following that, I shall look more closely at the reconstruction itself and then consider it within the context of Halley's own relationship to the past, his own stance in the battle of the books.

[18] A thorough account can be found in Fried and Unguru (2001).

Chapter 2
Apollonius's *Conics*

If there is some uncertainty about Halley's birth date, we are really in the dark about Apollonius's. The most one can say is that Apollonius of Perga was born around the middle of the third century B.C.E. and died sometime at the beginning of the second. Even this is only a matter of inference, based partly on a comment by Apollonius's late commentator and editor, Eutocius (c. 480–540 C.E.)—himself relying on yet another source, a certain Heraclius—that Apollonius was born "...in the time of Ptolemy Euergetes..." whose reign we know to have been from 246 to 221 B.C.E.[19] What can be said with confidence, however, is that Apollonius was much admired as a mathematician and that his great work was the *Conics.* This, at least, was the view of Apollonius by the end of the 1st century B.C.E., for Eutocius tells us, according to Geminus (1st century, B.C.E.),[20] it was "...on account of the remarkable theorems he proved about conics [that Apollonius] was called the Great Geometer."[21]

The contents of the *Conics* are set out by Apollonius himself in the prefatory letter for Book I. Apollonius tells us there that the entire work comprises eight books. Of these eight, the first four, he says, should be considered a "course in the elements" (*agōgē stoixeiōdē*) while the remaining four are "in the manner of additions" (*periousiastikōtera*), "special topics," one might say. Apollonius's overview of the books in his "course in the elements" is as follows: Book I contains the generation of the three conic sections and the opposite sections[22] and a "more complete and general" investigation of the characteristic properties, or *symptōmata,* of all these

[19] The comment by Eutocius is from his *Commentary on the Conics* (Heiberg, *Apolloniii Pergaei quae graece exstant cum commentariis antiquis, II*, p.168). More information may be found in Fried and Unguru (2001) and Toomer (1970).

[20] Little is known for sure about Geminus' life, and the time of his birth ranges from the 1st century B.C.E. to the 1st century C.E. In citing the earlier date, I am following Evans and Berggren (2006) here.

[21] Eutocius, *Commentary*, p.170.

[22] These are what we would call the two branches of the hyperbola. A hyperbola, for Apollonius, is only one of the branches. What this distinction says about Apollonius's general approach to the conics is discussed in Fried (2004).

M.N. Fried, *Edmond Halley's Reconstruction of the Lost Book of Apollonius's Conics*, Sources and Studies in the History of Mathematics and Physical Sciences, DOI 10.1007/978-1-4614-0146-9_2, sections[23]; Book II looks at the properties of diameters and axes and asymptotes, as well as "other things necessary and useful for determining limits of possibility" (*diorismous*)[24]; Book III includes theorems "useful for the synthesis of solid loci and for determining limits of possibility"; finally, Book IV concerns how many ways conic sections may meet one another and the circumference of a circle.

Except for the description of Book IV, which does give a fair picture of what that book is about, these descriptions are at best sketchy and give a far from complete account of Books I–III. Although Apollonius tells us that Book II looks at the properties of diameters and axes, for example, it is in Book I that many of the crucial theorems about diameters are proven—among others, that every conic section has an infinite number of diameters, that all diameters of the parabola are parallel, that all diameters of an ellipse or hyperbola are concurrent, that for every diameter of an ellipse or hyperbola there is a conjugate diameter and that the conjugate diameter of an ellipse is the mean proportional between the original diameter and its parameter, or *latus rectum* (a fact used over and over by Halley).

This is not to say that the descriptions of Books I–III are wholly uninformative as to their contents. Book I does contain much more than what Apollonius tells us, as I have just remarked; however, it truly does contain the generation of the sections and their *symptōmata*, which Apollonius chooses to emphasize in his prefatory letter. These things, moreover, are absolutely essential to Apollonius's treatment and conception of the conic sections, a conception that is a demonstrably geometric one, despite the ease with which key results in the *Conics* can be rewritten algebraically, at least by modern readers who are algebraically literate. To get a feeling for how Apollonius himself treats his subject, consider the statement and diagram for *Conics*, I.13, the proposition in which the ellipse is defined:

> If a cone is cut by a plane through its axis and is also cut by another plane which on the one hand meets both sides of the axial triangle and which on the other hand, when extended, is neither parallel to the base nor subcontrariwise, and if the plane containing the base of the cone and the cutting plane meet in a straight line perpendicular either to the base of the axial triangle or to it produced, then any [straight] line which is drawn—parallel to the common section of the [base and cutting] planes—from the section of the cone to the diameter of the section will equal in square some area applied to a straight line [the parameter] (to which the diameter of the section has the same ratio as the square on the straight line drawn—parallel to the section's diameter—from the cone's vertex to the triangle's base has to the rectangle which is contained by the straight lines cut off [on the base] by this straight line in the direction of the sides of the [axial] triangle), an area which has as breadth the straight line on the diameter from the section's vertex to where the diameter is cut off by the straight line from the section to the diameter and which area is deficient (*elleipon*) by a figure similar and similarly situated to the rectangle contained by the diameter and parameter. And let such a section be called an ellipse (*elleipsis*).[25]

---

[23] These *symptōmata* are given, in a somewhat simplified form, in appendix 1, as are some of the other basic Apollonian terms.

[24] The "limits of possibilities" are, roughly speaking, the conditions under which a problem has a solution at all or more than one solution. See below.

[25] I am taking advantage of the English translation by R. Catesby Taliaferro in Apollonius (1998).

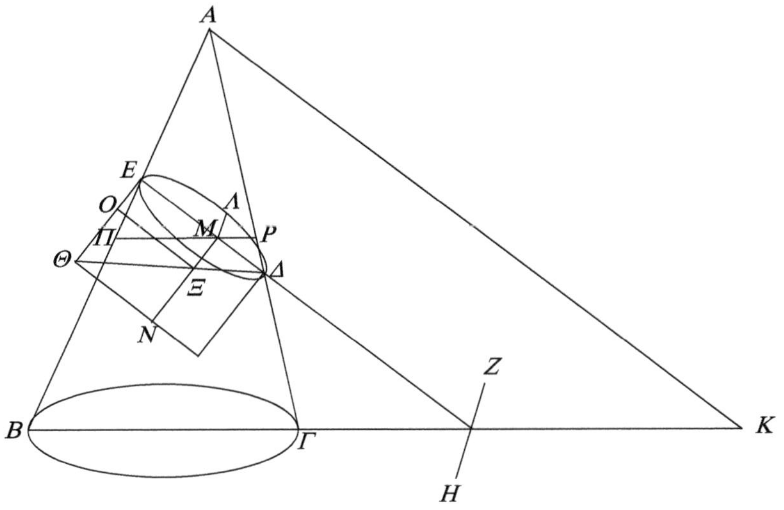

Fig. 2.1 *Conics* I.13

In fig. 2.1, then, a cone having base $B\Gamma$ and vertex A is first cut by a plane $AB\Gamma$ through the axis (i.e., the line from A to the center of the circular base $B\Gamma$), producing the "axial triangle" $AB\Gamma$; it is cut, next, by a second plane, $\Delta E\Lambda$, which meets two sides of the axial triangle and which, extended, intersects the plane of the base of the cone in a line HZ perpendicular to $B\Gamma$ (the line in which the plane $AB\Gamma$ meets the plane of the base of the cone). The section of the cone produced by the plane $\Delta E\Lambda$ is the ellipse. To define the *symptōma* of the ellipse we first need to determine the parameter $E\Theta$. It is the line that has the same ratio to ΔE as the square on AK has to the rectangle contained by BK and $K\Gamma$.[26] And with that, the *symptōma* is this: Let ΛM be drawn from the ellipse to the line ΔE, which is the diameter of the ellipse,[27] so that it is parallel to HZ. Then sq.ΛM = rect.$E\Theta,EM$ − rect.$O\Theta,O\Xi$, where the latter rectangle is similar and similarly situated to that contained by the parameter $E\Theta$ and the diameter ΔE.

The first thing one observes about this proposition is that it is long—impossibly long for most modern readers. One obvious reason for its length is that it is entirely stated in words: there are no symbols that formalize and condense statements and claims. After the enunciation, letters are introduced; however, they are not

[26] We shall use the shorthand, from now on, "sq.AK" for "the square on AK" and "rect.AK,KG" for the "rectangle contained by the sides AK,KG."

[27] That is, ΔE will bisect all chords of the ellipse drawn parallel to certain given line, in this case, line HZ. Earlier in Book I, in proposition I.7, Apollonius shows that any line such as ΔE, whether or not the plane $\Delta E\Lambda$ cuts sides AB and $A\Gamma$ of the axial triangle, will be a diameter in this sense. Also note that while HZ is perpendicular to $B\Gamma$ extended, it is not necessarily perpendicular to the plane of the axial triangle $AB\Gamma$, and, therefore, lines such as ΛM are not necessarily perpendicular to the diameter ΔE. If HZ is perpendicular to the plane $AB\Gamma$, then lines ΛM will be perpendicular to the diameter and the diameter will be an axis.

introduced as symbols, but only as pointers to parts of the diagram. Greek mathematics is tied to diagrams. An inseparable part of proposition I.13, therefore, is its diagram.[28] And so the enunciation is long because, more than a statement, it is a *description* that appeals to our ability to visualize geometrical operations, relations, and objects with the aid of diagrams. Anyone who reads Apollonius's own text closely must be left ineluctably with a sense of this geometrical mode of presentation, however one interprets it and whether or not one sees it also as Apollonius's mode of thought.[29] This sense of a geometrical presentation must be assumed regarding Halley, for even as a young man he knew the first four books of the *Conics* intimately,[30] and later, when he edited and translated the entire extant work, that intimate knowledge was extended to the remaining books, Books V–VII. Later, we shall see that Halley not only appreciated the geometrical form in which Apollonius presented his work, but also largely recognized that this was representative of a genuinely geometrical approach to the subject.

Books V–VIII, as we noted above, were included "in the manner of additions," according to Apollonius. It is not completely clear what Apollonius meant by this, but, minimally, he must have meant that these books were not conceived as general foundations on which a wide variety of investigations could be developed. It is in this sense that they can be said to differ from books of elements and reasonably referred to as books on "special topics."[31] Apollonius describes their contents in the letter introducing Book I as follows: Book V concerns minimum and maximum [lines], by which he means minimum and maximum of lines cut off between the axis of a section and the section itself[32]; Book VI treats questions of similarity and equality of conic sections—it is here, for example, that Apollonius proves that all parabolas are similar; Book VII contains theorems on the determination of limits of possibility *(peri dioristikōn theōrēmatōn)*; Book VIII, finally, Apollonius tells us, is a book of determinate conic problems *(problēmatōn kōnikōn diōrismenōn)*.

---

28 See Netz (1999).

29 Naturally, I have in mind here particularly H. G. Zeuthen's (1886), *Die Lehre von den Kegelschnitten im Altertum*, where Apollonius's geometrical presentation is thought to hide an algebraic mode of thought. Indeed, Zeuthen's point of view dominated the historiography of Greek mathematics at least until the 1970s; it was wholly adopted, for example, by Thomas L. Heath and Bartel L. van der Waerden. But besides abundant material in the *Conics*, and elsewhere in Greek mathematics, which is awkward to explain by means of Zeuthen's thesis, to say the least, it is dubious that Apollonius should have one mode of presentation and another, conceptually distinct, mode of thought. That said, a divide between presentation and thought is perfectly possible in the case of Halley, when he is occupied in the reconstruction of an ancient text.

30 Halley probably would have read Apollonius's *Conics*, I–IV from a text based on Federigo Commandino's famous edition of 1566. Commandino's edition included not only the first four books of the *Conics* but also Eutocius's commentary and Serenus's *On the Section of a Cylinder* and *On the Section of a Cone*. The latter were included with Halley's reconstruction of *Conics*, Book VIII.

31 Book IV also fits this description. In fact, its inclusion in the course of elements is not completely self-evident. The point is discussed at length in Fried and Unguru (2001), chapter III.

32 It has been often said that Book V concerns normals to conic section, but this characterization is moot.

More information about these later books of the *Conics* is found in the letters introducing each book individually. Particularly important for us is what is said in the letter introducing Book VII. For there, besides what is said in the introductory letter to Book I, Apollonius says the following:

> In this book [Book VII] are many wonderful and beautiful things on the topic of diameters and the figures constructed on them, set out in detail. All of this is of great use in many types of problems, and there is much need for it in the kind of problems which occur in conic sections which we mentioned, among those which will be discussed and proven in the eighth book of this treatise (which is the last book in it).[33]

So, here, unlike in the more general description in Book I, Apollonius explicitly connects the material of Book VII with the problems in Book VIII. As we have already seen, Apollonius's descriptions are not always completely adequate; however, the connection between Books VII and VIII is strengthened by the fact that, in Book VII of Pappus's (4th century C.E.) *Sunagōgē* or *Collection*, as it is usually known in English, lemmas to Books VII and VIII of the *Conics* are set out together, as if the two books were conceived as a pair. These hints about Book VIII in the *Conics* itself and in Pappus's *Collection* are important of course because Book VIII itself has been lost. We do not know for sure when it was lost, but by the 9th century, when great efforts were being made in the Islamic world to recover Greek mathematical and scientific texts, Book VIII was already missing even from the most complete manuscript of the *Conics* found in this period, the Banū Mūsā text.

---

[33] Toomer (1990), vol. I, p. 382. In general, all translations from the Arabic Books V–VII are from Toomer (1990).

Chapter 3
The Path to Halley

The Banū Mūsā text is central to our story, as it was Halley's principal source for his edition of Books V–VII and the basis of his reconstruction of Book VIII.[34] The manuscript was acquired sometime in the middle of the 9th century by the three sons of Mūsā b. Shākir, Muḥammad, Aḥmad, and al-Ḥasan, otherwise known simply as the Banū Mūsā, "The sons of Mūsā." Under the sponsorship of the Banū Mūsā, the text was translated into Arabic by Thābit b. Qurra (836–901).[35] The Greek original, subsequently, was lost, leaving to posterity only this Arabic version of Books V–VII. However, there was a long and crucial delay until even this Arabic translation of the later books of the *Conics* became fully available to mathematically minded readers in the Latin west. Of course the existence of Books V–VII was known from Pappus's account in the *Collection* and Apollonius's own account of them in his preface to Book I.[36] There were also attempts to reconstruct the contents of Books V–VII, notable of which was Vincenzo Viviani's reconstruction of Book V in 1659.[37] But before 1659, indeed sometime before 1629, the Banū Mūsā text was obtained by a Leiden mathematician, Jacobus Golius.[38] Golius, however, intending to produce his own translation from the Arabic, did not publish the text. And when Golius died in 1667, the manuscript, now in the hands of Golius's heirs, was still unpublished. Edward Bernard, who was soon to become Savilian professor of astronomy

[34] Another Arabic manuscript of the *Conics*, which was available to Halley, was that acquired by Christianus Ravius around 1640.

[35] There are references to another Arabic translation by a certain Isḥāq, but little is known of this. Toomer, however, says that Isḥāq's version might have contained part of Book VIII. See Toomer, *op. cit.*, p. xviii.

[36] Mathematicians were already acquainted with the early books of the *Conics* and Pappus's *Collection* as early as the thirteenth century (see Unguru, 1976 and 1974).

[37] *De Maximis et Minimis Geometrica Divinatio in Quintum Conicorum Apollonii Pergaei adhuc desideratum* (Florence, 1659).

[38] The story of Golius and the Banū Mūsā text is told in greater detail by Toomer (*op. cit.*, pp. xxii–xxv), and by Itard in his "L'angle de contingence chez Borelli," pp.113–117. Golius's manuscript of the Banū Mūsā text is now in the Bodleian Library, Oxford shelved as Marsh 667, after Narcissus Marsh who sponsored its purchase in Leiden in 1696 and who left it to the Bodleian Library as part of his bequest (see Toomer, *op. cit.*, pp. lxxxv–lxxxvii).

M.N. Fried, *Edmond Halley's Reconstruction of the Lost Book of Apollonius's Conics*, Sources and Studies in the History of Mathematics and Physical Sciences,
DOI 10.1007/978-1-4614-0146-9_3,

at Oxford (1673–1691), made a copy of the manuscript in 1668 and took some initial steps to translate the text. Finally, though, in 1696, Golius's heirs allowed the Banū Mūsā text and other manuscripts in Golius's collection to be auctioned in Leiden. From Leiden, largely through the efforts of Bernard, the manuscript eventually made its way to David Gregory, who replaced Bernard as Savilian professor of astronomy, and then to Edmond Halley, who was to become Savilian professor of geometry at Oxford in 1704.

The significance of the delay in transmitting the manuscript is clear. For while Apollonius's text lingered in the hands of Golius and his heirs, the rest of the mathematical world saw one of its richest periods of development. One need only think of such works as Descartes's *La Géométrie* in 1637, Desargues's *Bruillon projet* in 1639, Pascal's *Essay pour les Coniques* in 1640, Wallis's *Arithmetica infinitorum* in 1655, and, of course, Newton's *Principia* in 1687 to impress upon oneself the extent to which mathematics in that century was progressing in newly defined directions. Thus it is tempting to write, as Toomer has, that by the time Halley produced his edition of the Banū Mūsā text in 1710 "its interest was largely historical, whereas had it appeared before, instead of after, the work of the great 17th-century mathematicians such as Descartes, Fermat, Desargues and Newton (all of whom were thoroughly familiar with *Conics* I-IV), it could have influenced the development of mathematics in that century."[39] But to imply that, in this way, Halley's text marks a moment when the *Conics* became less important for what it might teach about conics *per se* than for what it might teach about the Greek *view* of conics brushes away the complexity in Halley's relationship to ancient texts such as this.[40]

One must remember, for one, that teaching Apollonius's *Conics*, together with Archimedes' works and all of Euclid, was still part of the official obligations of the Savilian geometry professorship in Oxford at the end of the 17th century.[41] Similarly, Savilian astronomy professors were to give due attention to classical texts; they were, for example, required to teach Ptolemy's *Almagest*. So when Henry Aldrich, Dean of Christ Church College Oxford, persuaded the newly elected Savilian professor of geometry, namely, Halley, to work with David Gregory on an edition of Apollonius, he was not only acting out of a keen interest in mathematics—Aldrich was a deeply interested and well-informed reader of mathematics, especially of classical mathematics[42]—but also out of an understanding of the requirements of Halley's position. Halley, for his part, took up the task with vigor, having already tackled a translation from the Arabic of another text by Apollonius, *The Cutting-off*

[39] Toomer, *op. cit.*, p. xxi.

[40] This point was missed, to some extent, in Fried and Unguru (2001).

[41] Cook (1998), pp. 323–324.

[42] See E. F. A. Suttle (1940). E. G. W. Bill (1988) also points out Aldrich's interest in classical mathematics and stresses that this was a pattern at Christ Church: "Perhaps above all the Deans of Christ Church encouraged the study of mathematics. John Fell, though not generally reputed a mathematician, contributed a preface in 1676 to John Wallis's edition of Archimedes and Eutocius in which he spoke of the disappointing reception of the proposal to issue an edition of *Veteres Mathematici*. His successor Aldrich was described when a tutor as 'a great mathematician of our house', and at the time of his death had an edition of Euclid in the press..." (pp. 268–269).

of a Ratio in 1706; Halley's enthusiasm, as has already been suggested and will be discussed further, was, like Aldrich's, not just a matter of zeal in fulfilling his responsibilities but also of genuine and profound interest. To characterize, then, what it might have meant to Halley to prepare an edition of Apollonius and attempt a reconstruction of Book VIII of the *Conics*, one must combine at least these three aspects of the context of his project: (1) the development in the 17th century of algebraic and other mathematical tools, allowing new and powerful means for investigating conics; (2) the requirements of the Savilian geometry professorship, which included attention to classical texts; (3) the recognition by Halley of the relevance and importance of these requirements and Halley's happily taking them up. More needs to be said especially about the third of these considerations. But before that, let us look a little more closely at how Halley approached the task of reconstructing the lost Book VIII of Apollonius's *Conics*.

Chapter 4
Halley's General Strategy for Reconstructing *Conics*, Book VIII

Whether or not it was deserved—and it may well have been deserved—Halley gives Aldrich the credit for pointing the way on how to reconstruct Book VIII. In the preface to the reconstruction, Halley recounts Aldrich's observation that "...in Pappus' *Mathematical Collection*, [Pappus] himself passed on lemmas serving [what] was to be demonstrated in the seventh and the eighth book of the *Conics* at the same time; whereas in the other books, different [lemmas] to the different [books] are provided. Hence, to you [Aldrich] we owe the discovery that the [two] books are conjoined; that the problems in [Book] eight are allotted their determination from the theorems on diorisms (διοριστικοῖς) in [Book] seven." Surely, by the time Halley wrote this, he would have known the connection between the contents of Books VII and VIII which Apollonius himself points out in the preface to Book VII, as described above. Still, Aldrich's observation regarding Pappus may truly have been the trigger that set the project going.

Somewhat disappointingly, though, when one actually looks at the lemmas for Books VII and VIII in Pappus's *Collection,* one finds they are not very helpful for surmising the contents of Book VIII. Of the fourteen lemmas assigned to Books VII and VIII, Hogendijk (1985, p. 45) rightly identifies eight that must refer to Book VIII alone. But as Jones (1986) concludes, these "... are so dismally elementary as to be almost completely worthless for the restoration of the lost material" (p. 498). And indeed Halley makes no use whatsoever of Pappus's lemmas for *Conics*, Book VIII, even where he might have done so, just to force the connection. In the alternate solution Halley suggested for VIII.15 and 16,[43] for instance, he might have cited Pappus's lemma 8 for *Conics* VII–VIII: "Let the sum of the squares on [two segments] AB and $B\Gamma$ be given and also the excess of the squares on AB and $B\Gamma$: [I say] that each of AB and $B\Gamma$ is given." Yet, Halley says nothing about Pappus here and relies completely on propositions 12 and 13 from *Conics* VII. It appears then that, for the most part, Pappus was important for Halley only in adding credence to the strategy of reconstructing Book VIII on the basis of Book VII, Halley's chosen course.

---

[43] At the end of VIII.16.

M.N. Fried, *Edmond Halley's Reconstruction of the Lost Book of Apollonius's Conics*, Sources and Studies in the History of Mathematics and Physical Sciences,
DOI 10.1007/978-1-4614-0146-9_4,

Without going so far as to say that Halley failed to appreciate Book VII of the *Conics* for its own worth, he did, nevertheless, treat it as a kind of table of contents for the problems of Book VIII, "everywhere observing the same order in which the diorisms are passed on [in Book VII]," as he says in his conclusion. Table 1, showing the dependence of propositions in Halley's text (indicated with a capital H) on propositions from Book VII, makes this abundantly clear. From the proposition numbers alone, one can see that, except for four propositions, namely, *Conics* VII.1, 4, 32, 44,[44] Halley covers all of Book VII in the course of his reconstruction. But not only does Halley try to exhaust the contents of Book VII, he also shadows its thematic structure. Propositions in the first half of *Conics*, Book VII can, to a great extent, be paired with *diorisms*, "limits of possibility," in the second half. In his reconstruction of Book VIII, then, Halley typically formulates his problems on the basis of these pairs, taken one after the other, so that, in this way, his reconstructed Book VIII advances *pari passu* with *Conics*, Book VII.

Table 1. The Dependence of Halley's Book VIII on *Conics*, Book VII

H1: VII 5	H12: VII 9 12 27	H23: VII 16 6 (as stated by Halley; 36 is meant) 29
H2: VII 5 32	H13: VII 10	H24: VII 16 (29) (30) 37
H3: VII 29	H14: VII 10 12 28	H25: VII 17 38 39 40
H4: VII 30	H15: VII 11 (31)	H26: VII 17 30 41
H5: VII 2 6 13 (28)* 29	H16: VII 12 13 14	H27: VII 18 42
H6: VII 3 7 12 30	H17: VII 31	H28: VII 18 29 30 43
H7: VII 6 13 21 22 23 29	H18: VII (12) 31	H29: VII (6) 19 41
H8: VII 7 24 12 30	H19: VII (6) 15 29 33 34 35	H30: VII 19 23 (29) 45 46
H9: VII 8 13 25 29	H20: VII 15	H31: VII 19 47 48
H10: VII 8 26 30	H21: VII 6 (21) (22)	H32: VII 20 49 50
H11: VII 9 13 20 27	H22: VII 7 (31)	H33: VII 20 51

*Parentheses mean the proposition is only implied.

Thus, for example, having completed problem VIII.8, Halley has taken care of the problem of constructing conjugate diameters having a given ratio. This is a problem inspired by *Conics* VII.7 and its related diorism in *Conics* VII.24, both

44 Two of these propositions, VII.1,32, concern the parabola, which was of little interest to Apollonius in this book, except perhaps for setting up analogies (see Fried (2003)). The parabola also, correspondingly, plays a minor role in Halley's reconstruction.

of which concern the ratio of conjugate diameters.[45] Halley then continues VIII.9, which states:

> Given the axis and latus rectum of a hyperbola, find conjugate diameters of it, both in position and magnitude, whose sum equals a given line.

This is the situation referred to by *Conics* VII.8 (stated here in terms of the letters in fig. 2, as it is in the *Conics* itself):

> ... [in a hyperbola or ellipse] the ratio of the square on $A\Gamma$ (which is the transverse diameter) to the square on BK and ZH, the two conjugate [diameters], when they are joined together in a straight line [i.e., the square on the sum of the conjugate diameters], is equal to the ratio of $(N\Gamma \cdot M\Xi)$ to the square on a line equal to line $M\Xi$ plus the line which is equal in square to $(MN \cdot M\Xi)$.

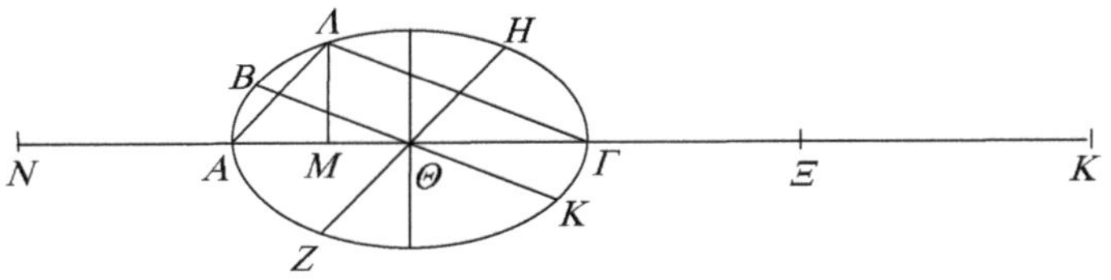

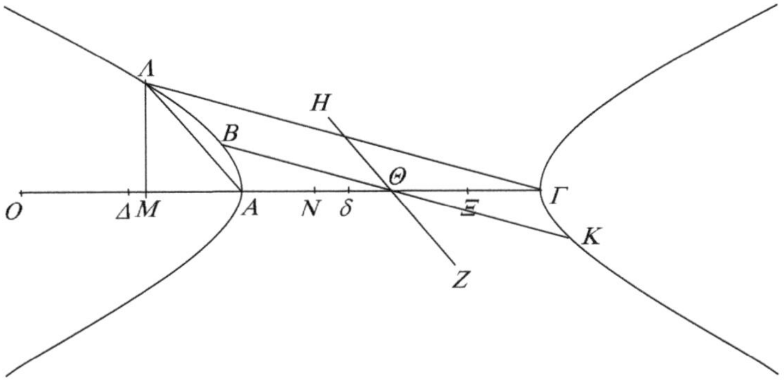

Fig. 2. Halley VIII. 9.10 *Conics* VII.8

The diorism related to this proposition is *Conics* VII.25, which gives a lower bound for the sum of the conjugate diameters:

> [In] every hyperbola, the line equal to [the sum of] its two axes is less than the line equal to [the sum of] any other pair whatever of its conjugate diameters; and the line equal to the sum of a transverse diameter closer to the greater axis plus its conjugate diameter is less than the line equal to the sum of a transverse diameter farther from the greater axis plus its conjugate diameter.

[45] See the main text of the translation for the statements of these.

And this, accordingly, Halley cites at the end of the problem. VIII.10, subsequently, takes up the same problem for the ellipse also treated by *Conics* VII.8 (fig. 2), and draws from the next diorism, VII.26; problem VIII.11 is based on *Conics* VII.9 and diorism VII.27. And so it goes: for each proposition and diorism in *Conics*, Book VII, a problem or set of problems in the reconstructed Book VIII; the next proposition, the next problem.

This pattern dominates Halley's reconstruction; however, it is not followed mindlessly and slavishly. One revealing break from it is Halley's VIII.17 (continued in VIII.18). Problem VIII.17 sets the task of finding the magnitude and position of conjugate diameters containing a given angle, given the axis and latus rectum of a hyperbola. While this problem is a natural continuation of the previous series of problems focused on the construction of conjugate diameters, it has no direct antecedent in *Conics*, Book VII. In a rare instance of explicit justification, Halley writes immediately following the enunciation of problem 17:

> The solution to this problem, as well as to the next, is provided by Apollonius at the end of Book II: there, however, he assumes the sections are already described. Among *diorism* theorems in the seventh book, proposition 31 seems to have been inserted to pave the way for this very problem when it has been assumed that the curve has not yet been described, as stated in the premises.

The reference to *Conics*, Book II is almost certainly to II.51 in which Apollonius shows how to find a tangent to a given ellipse or hyperbola so that the tangent has a given angle with the diameter passing through the point of tangency. Since the conjugate diameter is parallel to the tangent and bisects the original diameter, its position is given, and since the sections themselves are given, its magnitude can be found as well. Halley then shows that with VII.31, a beautiful proposition showing that in an ellipse or opposite sections any parallelogram formed by the tangents at the vertices of conjugate diameters is equal to the rectangle formed by the tangents at the vertices of conjugate axes (see fig. 3), knowledge of the axis and latus rectum alone is sufficient to find the required conjugate diameters. More specifically, knowing the axis and its latus rectum allows us to know also the conjugate axis, and, therefore, the rectangle contained by the axes. This rectangle is equal to the parallelogram on the conjugate diameters by *Conics* VII.31. And since the angle between the diameters is given, the rectangle having sides equal to the conjugate diameters is also given. Thus the problem is reduced to one of finding the position and magnitudes of conjugate diameters containing a given rectangle, which Halley already solved in his VIII.13.

For the present discussion, however, the mathematical argument is less important to us than Halley's explicit appeal to Apollonius's own development of Book VII. True to his basic assumption that Book VII provides the material for the problems in Book VIII, Halley persistently asks how propositions there, especially striking ones such as VII.31, might have grounded problems set out by Apollonius. Halley's approach then involved his putting himself aside and trying to enter Apollonius's thinking and ways of working. This is visible to an extent even in Halley's Latin, which sometimes becomes awkward in its attempt to follow Apollonius's Greek mathematical style. For instance, when Halley wants to say that the square on a line

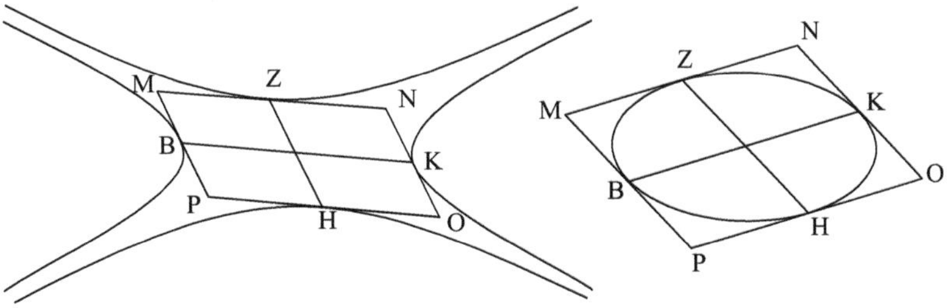

Fig. 3. *Conics* VII. 31: Parallelograms MNOP formed by the tangents at the vertices of conjugate diameters always have the same area

is equal to, say, the difference between the square on the axis and its figure, as in VIII.5, he writes: "*...quae...poterit differentiam quadrati ex Axe et figurae ejusdem...*[Halley's capitalization]." This strange usage of the verb *posse*, "to be able," can only be understood in light of the Greek verb, *dunasthai*, whose idiomatic meaning in Greek mathematics, including Apollonius's mathematics, is "to be equal in square."[46] Naturally, the totality of Halley's text speaks louder than these linguistic matters; nevertheless, they are indicative of the seriousness with which Halley adopted his particular modus operandi vis-à-vis Apollonius's lost work.

[46] I owe thanks to Marinus Taisbak for clarification on this point. Of course Halley may not have been the very first to try to catch the Greek idiom in this way, and, certainly, his usage was consistent with translations such as this by Commandino of Euclid's *Elements,* Book X, definition 3: "Rectae lineae *potentiā* commensurabiles sunt..." for *Eutheiai dunamei summetroi eisin...* ("Straight lines are commensurable *in square...*"). Still, it shows the care he took to be as Greek as he could be.

Chapter 5
Halley's Dialogue with the Past

Given Halley's attentiveness to Apollonius's text, his attentiveness to those "certain concrete vestiges," as he puts it in his epistolary introduction, which hint at Apollonius's thought regarding the book of determinate problems, given his apparent desire to enter Apollonius's mind, Halley's project, contrary to what was said above, begins to look purely historical after all. And one could even accept this identity as a historian without denying Halley's identity as a mathematician. For, certainly, while his reconstruction demanded a considerable degree of mathematical ingenuity, such as one might expect from the new Savilian professor of geometry, one need not assume the reconstruction was only a pretense for Halley to exhibit his mathematical prowess or to explore new mathematical themes, Apollonian or not, that might flow from *Conics*, Book VII. Yet the fact remains that Halley *did* come to the project as a mathematician and scientist and did so willingly, as I have already stressed. So the truth is, while Halley's project should not be viewed as purely mathematical, it should not be viewed as purely historical either: we must look for the character of Halley's relationship with texts of the past, like Apollonius's *Conics*, in the middle ground between historical sensitivity and rigor and mathematical insight and interest.

Halley's place in this middle ground will be better understood if we consider his own reconstruction of *Conics*, Book VIII in the light of other reconstructions of Greek mathematical works. Indeed one ought to be well aware that Halley's attempt to reconstruct such texts was hardly an unprecedented enterprise. We have already mentioned Viviani's 1659 reconstruction of Book V of the *Conics*, and Francesco Maurolico (1494–1575), in the previous century, attempted a reconstruction of Books V and VI of the *Conics*. This reconstruction activity, however, accelerated significantly in the 17th century. That century opened with François Viète (1540–1603) producing his reconstruction of Apollonius's *On Tangencies*, his *Apollonius Gallus* of 1600. Subsequently Marino Ghetaldi (1568–1626) published a reconstruction of Apollonius's *Neusis* (1607) and Wilebrord Snell (1580–1626) reconstructed the *Determinate Section* in his *Apollonius Batavus* (1608), being the Dutch equivalent to Viète's "French Apollonius." This activity, moreover, continued, though with somewhat less intensity, well into the 18th century and even into

M.N. Fried, *Edmond Halley's Reconstruction of the Lost Book of Apollonius's Conics*, Sources and Studies in the History of Mathematics and Physical Sciences,
DOI 10.1007/978-1-4614-0146-9_5,

the 19th century. Thus, for example, one has Robert Simson's (1687–1768) reconstruction of Apollonius's *Plane Loci* in 1749. And Michel Chasles (1793–1880) reconstructed Euclid's *Three Books of Porisms* in 1860. Significant for us is that in his introduction to his *Porisms* Chasles mentions Halley and his reconstructions, including that of *Conics*, Book VIII:

> The celebrated astronomer Halley, being well-versed in the geometry of the Greeks, translated from Arabic the treatise *Cutting-off of a Ratio* and restored *Cutting-off of an Area* and the 8th book of Apollonius's *Conics*, as is well known. The enigma of the *Porisms* must have indeed attracted him.... (p. 5)[47]

On the same page, Chasles also recalls Viète's and Snell's reconstructions. Chasles's introduction thus underlines Halley's place in a tradition of restoring Greek mathematical works, particularly in its 17th-century heyday.

On the other hand, the motives behind reconstructions in the first half of the 17th century, such as those of Viète, Ghetaldi, and Snell, were directed in a way that they were not by the end of that century and the beginning of the new one. These early 17th-century mathematicians approached their task feeling that they now possessed a key for understanding their ancient predecessors and for reconstructing their thought. The key was the analytic art, as Viète called it, or algebra. For this reason they chose works such as Apollonius's *Plane Loci* or Euclid's *Porisms* that Pappus included in the corpus of texts based on analysis: reconstructing these works, therefore, became for them a way of working out the potential of the new algebra. This is especially clear in the case of Fermat's reconstruction of Apollonius's *Plane Loci*, which he worked on from about 1628 to 1636. For what followed on the heels of that reconstruction was his more famous *Introduction to Plane and Solid Loci* (*Ad locos planos et solidos isagoge*), famous because it contained the most complete presentation of Fermat's version of analytic geometry. Michael Mahoney, who gave a thorough and deep account of the connection between these two works (Mahoney, 1994), summed up the situation as follows:

> Fermat was no antiquarian interested in a faithful reproduction of Apollonius' original work; he was a working mathematician seeking to ferret out the analytic techniques he felt Apollonius had hidden. The *Plane Loci* was to serve as a means to an end rather than an end in itself. (p. 96)

In this sense, reconstruction at the start of the 17th century had a direct effect on the development of the mathematics of the period. By the end of the century, however, when Halley had begun his own reconstructions, algebra was no longer an art whose worth needed justifying. So although Halley's choice of texts was squarely within the tradition established in the 17th century—and here one must remember that the *Conics* was among the principal works Pappus assigned to the domain of analysis—his motivations must have been much more nuanced than those of Viète and Fermat.

---

[47] "Le célèbre astronome Halley, très-versé dans la connaissance de la geométrie des Grecs, traduisit de l'arabe, comme on sait, le Traité de la *Section de raison*, at rétablit celui de la *Section de l'espace* et le VIIIe livre des *Coniques* d'Apollonius. L'énigme des Porismes devait naturellement lui offrir de l'attrait."

It is perhaps easier to discern the character of Halley's stance toward his own reconstruction activity, and specifically regarding Book VIII, if we contrast it with the other great attempt to restore Book VIII of the *Conics*, namely, Ibn al-Haytham's early 11th century *Treatise on the Completion of the Conics.*[48] On one level, Ibn al-Haytham, or Alhacen as he became known in Europe,[49] pursued his task in a way that was similar to Halley's. Like Halley, Ibn al-Haytham also wanted to keep the *Conics* and Apollonius always in view. Several of the problems in Ibn al-Haytham's reconstruction too are almost identical to those of Halley.[50] For instance, Ibn al-Haytham's problem set[51] 8 requires finding a diameter of a hyperbola or ellipse which together with its latus rectum contains a given rectangle. This corresponds to Halley's problems 27 and 28: "Given the sides of the figure of the axis of a hyperbola [prob. 27; "ellipse" in prob. 28], it is required to find the position of a diameter that has its figure, or rectangle contained by the diameter and its latus rectum, equal to a given rectangle." The task in Ibn al-Haytham's problem set 9 is to find a diameter of a hyperbola or ellipse which when added to its latus rectum is equal to a given sum. This corresponds to Halley's problems 25 and 26: "Given the axis in a hyperbola [prob. 25; "ellipse" in prob. 26] and its latus rectum, it is required to find the position of that diameter which together with its latus rectum comes to a given sum." Again, problem set 10 in Ibn al-Haytham's work asks for a diameter of a hyperbola or ellipse that has a given ratio with its latus rectum, exactly what is required in Halley's problems 21 and 22: "Given the axis of a hyperbola [21; "ellipse" in 22] and the latus rectum of the axis, find a diameter which has to it latus rectum a given ratio."

On the other hand, although the problems just cited were based on Book VII, Ibn al-Haytham did not restrict himself to Book VII, and he set out significant problems which could be related to the *Conics* in a much less direct way. Problem set 4, for instance, requires constructing a tangent to a conic section that intersects a given tangent so that the ratio between the two segments defined by the points of tangency and the point of intersection is equal to a given ratio; problem set 11, to take another example, is a *neusis*-like problem requiring the construction of a line passing through a given point and cutting a segment of a conic section so that the part of the line cut off by the segment is equal to a given line. These are far from trivial problems, and Ibn al-Haytham would have viewed their solution as an achievement.

Problems like these were included in al-Haytham's *Completion of the Conics* because, as he puts it in his introduction, "When we studied this work [the *Conics*], investigated the notions in it, and went through the seven books many times, we found

---

48 *Maqāla fī tamām kitāb al-makhrūṭāt.*

49 His full name was Abū Ali Al-Ḥasan ibn al-Ḥasan ibn al-Haytham. "Alhacen" is apparently a corruption of Al-Ḥasan.

50 There is no reason to believe Halley knew of Ibn al-Haytham's work and very good reason to believe that he did not. At present, only a single manuscript of Ibn al-Haytham's *Completion of the Conics* exists in Manisa, Turkey, and that came to light only around 1970.

51 I am using Hogendijk's (1985) groupings here: each grouping represents a series of thematically related problems. Hogendijk divides the book into 11 such problem sets.

that it lacked notions, which this work *should not leave untreated* [my emphasis]."[52] Such things, in other words, were worthy of Apollonius, and since they did not appear in the extant books of the *Conics,* it was likely that they appeared in the lost book. And Ibn al-Haytham could consider himself able to judge what was worthy of Apollonius because he saw himself as a fellow mathematician and one whose own thoughts about conics were consistent with any of those Apollonius might have thought. As Hogendijk (1985) argues, "Ibn al-Haytham's supposition [regarding the "notions" he judged to have been necessarily included in Book VIII] is that Apollonius gave a complete treatment of certain classes of related problems. It seems to me that he based this supposition not on evidence in the *Conics*, but only on his implicit assumption that Apollonius's interests were identical to his own" (p. 69). In other words, in Ibn al-Haytham's view, Apollonius's and his own mathematical thought were coterminous, and, therefore, he could see himself as continuing Apollonius's work, rather than as trying to imagine, as a historian would, how Apollonius might have continued his work.

With Ibn al-Haytham's work in mind, one can begin to discern the subtle but crucial difference between a "completion," as Ibn al-Haytham called his work, and a "restoration," as Halley called his. More pointedly, in the difference between "completion" and "restoration," one can begin to see two different attitudes toward mathematics of the past.[53] With the former, while one may be looking back at a work of the past, one uses it with an eye to producing a work on the frontier of new knowledge; the antiquity of the original work is an almost incidental matter, except, perhaps, that one's present work *might* have been done previously before being lost; one might as well refer to the original work as that of a brilliant colleague who has died and whose work was lost in a fire. With a restoration, on the other hand, there is an awareness of an essential difference between the past and present and, accordingly, a sense that one ought to be faithful to the original mathematical text. Of course how far one recognizes the difference between past and present, ancient and modern, is not a minor detail. A modern historian of mathematics takes the difference as absolute: the idiosyncratic approaches and understandings of mathematicians of the past are, therefore, at the center of the historian's investigation. Unguru (1979) has written, accordingly:

> History, as Aristotle knew, focuses on the idiosyncratic rather than the nomothetic. It is impossible for modern man to think like an ancient Greek. Historical understanding, however, involves the attempt at faithful reconstruction of the past. In intellectual history this necessarily means the avoidance of conceptual pitfalls and interpretive anachronisms. (p. 556)

By contrast, we have approaches to historical mathematical texts via modern mathematics, as in the case of Zeuthen's (1886) treatment of Apollonius.[54] Historians like Zeuthen may still recognize a difference between ancient and modern, at least in modern mathematics's possession of powerful tools lacking in theancient

[52] Translation from Hogendijk (1985), p. 134.

[53] The existence of this difference makes the question of which reconstruction of Book VIII is the truer reconstruction somewhat moot.

[54] See note 28 above.

world, but they tend to adopt a position similar to the one Hogendijk attributed to Ibn al-Haytham, namely, that the interests of mathematicians like Apollonius were "identical to [their] own."

If we had to place Halley somewhere between a mathematician historian like Zeuthen and a modern historian of mathematics, we should probably place him closer to the former than to the latter. Like the mathematician historians, Halley, as a working mathematician and scientist, might well have judged himself to be in a privileged position for interpreting the works of mathematicians of older times.[55] And like them, he was in possession of powerful tools, in particular algebraic tools, absent from Apollonius's arsenal.[56] As we have already remarked, Halley's life began in a mathematical world informed by Descartes, Desargues, Pascal, and Fermat and flowered in one inhabited by Wallis, Newton, and Leibniz. On the other hand, while modern mathematicians are fully cognizant of how far algebra and analysis can take one in exploring the properties of conic sections and other kinds of mathematical curves, mathematicians at the end of the 17th century could still debate these advantages and could still use geometrical techniques without apologizing for the approach or treating them as *merely* historical[57]: they were not as close to the ancient mathematicians as Ibn al-Haytham, but they were not as far from them as post-18th-century mathematicians were. Indeed, solving a problem or proving a theorem in the style of the ancients could be taken as a mathematical challenge as serious as any other. In the *Principia*, for example, Newton takes up the problem of the four-line locus,[58] and at the end of the second corollary to Lemma XIX (Book I, part V), he writes: "And so we have given in this Corollary a solution of that famous Problem of the ancients concerning four lines, begun by *Euclid*, and carried on by *Apollonius*; and this not an analytical calculus but a geometrical composition, such as the ancients required."[59]

So how should we define Halley's relationship to the past and, hence, his endeavors to reconstruct Apollonius's lost works? From what has been discussed so far, it is evidently not one in which Halley sees himself continuing Apollonius's

---

[55] For this aspect of Zeuthen's historical outlook, see Lützen and Purkert (1994).

[56] Halley was well aware of the advantages these afforded him. An explicit expression of this can be found, for example, in the opening paragraph of Halley's (1694) "Method of finding the Roots of Aequations Arithmetically," which was appended to Newton's *Universal Arithmetick*: "The principal Use of the Analytick Art, is to bring Mathematical Problems to Aequations, and to exhibit those Aequations in the most simple Terms that can be...The Antients scarce knew any Thing in these Matters beyond Quadratick Aequations. And what they writ of the Geometrick Construction of solid Problems, by the Help of the Parabola, Cissoid, or any other Curve, were only particular Things design'd for some particular Cases. But as to Numerical Extraction, there is every where a profound silence; so that whatever we perform now in this Kind, is entirely owning to the Inventions of the Moderns" (translation by Joseph Raphson and corrections by Samuel Cunn, London, 1720).

[57] See Boyer (1956), especially, chapters VI and VII.

[58] The problem is this. Suppose four lines are given fixed in position and that from a point lines are drawn at given angles to each of these four given lines. Then the locus of points for which the rectangle formed by two of the drawn lines *has a given ratio* to the rectangle formed by the other two lines will be a conic section given in position.

[59] From vol. 1, p. 81 of Newton (1962).

investigations, one in which he considers himself a mathematician of the same ilk as the ancients, though, by unfortunate circumstances, living in a time inferior in knowledge; Halley's relationship to the past, in short, is not like that of Ibn al-Haytham. But nor is it one in which he sees himself occupying a position of superior knowledge where he could feel assured that, from a purely mathematical point of view, Apollonius's investigations would be pursued more neatly and explained more broadly with modern techniques and conceptions, where looking at Apollonius is a fascinating, but at bottom antiquarian pursuit, an archaeological pastime.[60] On the contrary, Halley was one of those 17th-century mathematicians alluded to above who did not immediately treat the mathematics of the ancients as hopelessly inferior to that of the moderns, even while he appreciated the modern conceptions and techniques profoundly and excelled in them as well as anyone of his age. It is in this spirit, for example, that he composed the preface to his earlier translation of Apollonius's *Cutting-off of a Ratio* and reconstruction of *Cutting-off of an Area* (1706), which he opened by extolling the modern achievement of the "Algebra of Species," the "Arithmetic of Infinitesimals," and the "Fluxions," referring to the works of Viète, Wallis, and Newton, but then continued by urging that this should not in any way lessen the glory of the ancients who brought geometry to perfection[61]; indeed, later in the preface, while he recognizes how algebra is applicable to Apollonius's work, he is at the same time quite critical of Descartes, especially regarding Descartes's claim that Apollonius could not fully solve the four-line locus problem.[62] So Halley stands both mindful of his own powerful mathematical contemporaries and yet also respectful of mathematicians of the past, like Apollonius, whose methods and thinking he judged worthy of admiration. I think the best way to describe the nature of his relationship to ancient mathematical thinkers is that it was the nature of a dialogue. By translating and editing existing mathematical works of Apollonius and reconstructing the lost works, Halley was engaging Apollonius in a kind of conversation, such as Halley relished at Jonathan's, Garraway's, and other coffee houses he frequented.

In the reconstruction of Book VIII, this sense of dialogue can be felt not only in its explicit appeals to Apollonius, described above, but also in frequent additions to problems that are unmistakably in Halley's own voice. These additions, which are either alternative solutions to the problems or variations on them, can be found in Halley's VIII.5, 6, 7, 8, 9, 10, 11, 12, 14, 23, 24, 26. Often these are introduced by saying the problem can be solved "not awkwardly" (*ne inconcinne*) (VIII.9), or "most expediently" (*satis expedio*) (VIII.11), or "in a not at all inelegant way" (*modo sane non ineleganti*) (VIII.12), or "in an entirely different way"

[60] I do not mean archaeological in the Foucauldian sense, but in the somewhat pejorative sense of digging up old bones from dead ages.

[61] "Quamvis de scientiis Mathematicis, hac nostra et superiore aetate, praeclare meruerint Viri eruditi, qui Algebram Speciosam, Arithmeticam Infinitorum, nuperamque Fluxionum doctrinam adinvenerunt et excoluerunt: nihil tamen inde Veterum gloriae detrahitur, qui Geometriam ad eam provexere perfectionem..."

[62] Halley refers at this point to the set of lemmas in Newton's *Principia* from Part V of Book I, which I mentioned above.

(*modo prorsus diverso*) (VIII.14). Halley's intention to distinguish, with some pride no doubt, his own contributions from Apollonius's reconstructed solutions is patent in these phrases. Yet, this is not Halley playing a game of one-upmanship: his contributions are meant only to be a different view of the matters at hand.

Thus, nearly all Halley's additions are substantively different from Apollonius's reconstructed solutions. They differ, specifically, in their attention to magnitudes only, for example to the magnitudes of diameters, as opposed to their magnitudes *and* positions. For example, Apollonius's problem 7 in Halley's reconstruction reads:

> Given the axis and the latus rectum of the axis of a hyperbola, and given the ratio of conjugate diameters of the section, find the conjugate diameters *both in magnitude and in position* [my emphasis].

But then, having presented his proposal for Apollonius's solution, Halley adds:

> Since the difference between the squares on the conjugate diameters is always equal to the difference between the squares on the axes, though, we can give this solution to the problem in a fairly expedient way, *but without the position of the diameters* [my emphasis].

Halley's construction, as he says, is quite simple.[63] Let the given ratio be $P\Sigma$ to ΣT. Draw the arc of a circle with center Σ and radius $P\Sigma$. Draw TX so that it is perpendicular to ΣP, and let it intersect the circle at X. Let $T\psi$ be cut off of TX so that the square on $T\psi$ is equal to the difference of the squares on the axes of the section,[64] and, therefore, also equal to the difference of the squares of the required diameters.

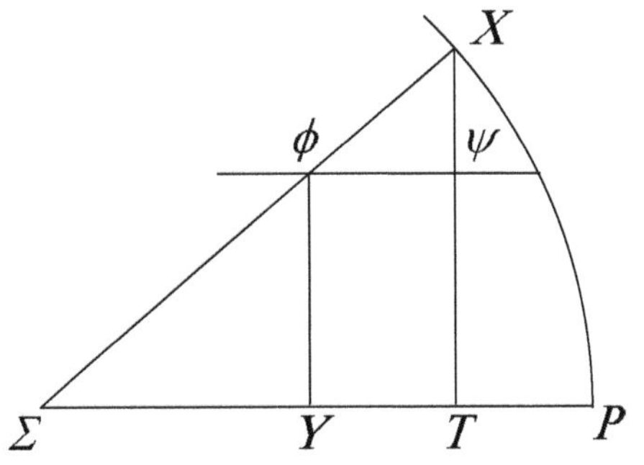

Conics VII.13. Draw $\psi\phi$ parallel to ΣT, so that it intersects ΣX at ϕ. Then $\Sigma\phi$ and ΣY are the required diameters, for the difference of their squares is equal to the square on $Y\phi$, which is equal to the difference of the squares of the required diameters, and the ratio of $\Sigma\phi$ to ΣY is the same as the ratio of ΣX to ΣT, that is, ΣP to ΣT.

[63] I am leaving out some details, which can be found in the translation itself.

[64] By *Conics* I.16 and the definitions following it, this is equal to the rectangle contained by the axis and the difference of the axis and its latus rectum, all of which are given since the axis and latus rectum are given.

The solution itself is geometrical, but the geometry has little to do with the specific geometry of the hyperbola, only with certain given relationships of magnitudes: it is a solution of the abstract problem of finding two magnitudes that are in a given ratio and whose squares have a given difference. If this exclusive focus on magnitudes here and in Halley's other additions does not actually arise out of a Cartesian tendency (despite Halley's own occasional objections to Descartes!)[65] to see the essence of problems in algebraic or numerical relationships, it nevertheless does lend itself easily to that kind of algebraic development. Halley was not averse to algebra, and there are places in the restoration where an algebraic mode of thought is apparent.[66] It is certainly the case that numerical relationships play a part in his thinking even in this seemingly nonnumerical work, as can be seen in problems 30 and 31, where there appear numbers such as $\sqrt{2}+1$ and $\sqrt{8}+2$ and even decimal approximations such as 0.6436 for $\sqrt{(\sqrt{2}-1)}$ (in problem 31). These numerical results are surely not the "Great Geometer" speaking, but Halley, who loves to measure and calculate, speaking to him.

It is worth noting too that, as one might guess by the remarks referring to the preface of *Cutting-off of a Ratio* (1706) above, Halley's approach to the *Conics* in 1710 was not significantly different from his approach in that earlier work of 1706. In the *Cutting-off of a Ratio*, naturally, the additions are set off from the text as *scholia* rather than incorporated directly into the text. And these *scholia* also bring to the text both alternative solutions to the problems in the work as well as more modern, algebraically expressed considerations, as for example in the *general scholium* ending Book I of *Cutting-off of a Ratio*. In the reconstruction of *Cutting-off of an Area*, which accompanied *Cutting-off of a Ratio*, one sees again the appeal to Apollonius throughout the work; in fact, it often gives the impression of being more an extended discussion of the work than a full-fledged reconstruction. But perhaps none of this is surprising given what we have seen of Halley's general outlook toward the past.[67]

To sum up then, the metaphor of a dialogue catches two central and closely related aspects of Halley's attitude toward the past. The first is that he genuinely respected ancient thinkers and their work. He neither spoke down to them nor up to them; he saw them as intelligent interlocutors. And second, he felt he could learn from engaging with the past. Halley learned from the past in more than one way. One of these involved using the past as a source of data, as in Halley's use of Hipparchus's data[68] as recorded in Ptolemy to determine the proper motion of stars

---

[65] See above.

[66] See, for example, the note 153 problem 13 in the translation.

[67] I might emphasize, however, that Halley's consistency of approach is particularly striking given the fact that when he worked on the *Cutting-off of a Ratio* and the *Cutting-off of an Area*, he was still working closely with David Gregory. But David Gregory died in 1708. So Halley's work on the extant books of the *Conics* and the reconstruction of Book VIII was done largely on his own. This suggests that Gregory's influence on Halley's general approach to the Greek mathematical works was either minimal or so complete that Halley continued in the same vein after Gregory's death. The latter does not seem very likely. Halley's approach to the works of Apollonius then was truly his own, even if it was in perfect sympathy with Gregory's.

[68] In particular, concerning the positions of the bright stars Aldebaran, Sirius, and Arcturus.

and of course in his cometary work. Even here, though, the seriousness with which Halley used historical data shows the seriousness with which he viewed the collectors of those data.

But the past, for Halley, was not only a repository of data or, as it might have been for an earlier figure such as Ibn al-Haytham, a source of lost knowledge: it was also a font of intelligent treatments of problems and ideas by thoughtful people. By paying attention to mathematicians and astronomers of the past, he could improve himself both by their account of the world and also by their ways of approaching questions and their ways of viewing the world; this historical horizon could broaden his own horizon. For this reason, regarding Halley's historical interests in science, Allan Chapman (1994) has written that, "it was when [Halley] came to view research into nature itself that we see his use of history most creatively displayed: for what interested Halley most of all was understanding the processes by which nature worked, and this included understanding how those processes were perceived by previous generations of mankind" (pp. 167–168).

It may be because of Halley's ability to listen to other voices from other times and to appreciate how even an ancient treatment of a problem may be useful and enlightening that he was such a good mediator for science. An instance of mediation pertinent for the present discussion, if only as an image, was Halley's visit, at the behest of the Royal Society, to the astronomer Johann Hevelius of Danzig in 1679. At issue was a dispute that had begun some years before, mainly between Robert Hooke and Hevelius, in which Hooke had criticized Hevelius severely for his use of open[69] instead of telescopic sights in measuring star positions. Halley, having established his reputation as an observational astronomer in St. Helena, was sent to observe Hevelius's methods. Although Halley himself used the modern telescopic sights in his own astronomical work, he learned Hevelius's procedures, appreciated them, and, ultimately, defended Hevelius.[70]

In a way, albeit a qualified way, Halley acted in this dispute as an ambassador from the new to the old. In the end of course the new would win out, and yet Halley could give a fair chance to the old and learn from it. To be a good mediator means to be an agent of dialogue between the sides, to look closely at both, and represent one to the other faithfully and fairly. It is in this role that we should understand Halley's work on ancient mathematical texts in general and on his reconstruction of Book VIII in particular. He tried to see such texts as they were and to appreciate them as they were, but not with this as his single uncompromising goal, as it would be for a historian. He came to such texts as a modern, as any modern would; yet, he did not try to find in them modern texts in disguise. He was, as I have said, a mediator between the ancient and the modern mathematical world, giving the former all its historical mathematical due.

---

[69] That is, a sight used with the naked eye.

[70] Later Halley's relations with Hevelius soured somewhat as did Halley's enthusiasm for Hevelius's techniques. See Ronan (1969), chap. 4, and Cook (1998), chap. 4 for detailed accounts of this chapter in Halley's career. Some of Halley's correspondence with Hevelius can be found in MacPike (1932).

Chapter 6
A Note on the Translation

Someone remarked to me once that Ver Eecke's French translations of Apollonius, Pappus, and Diophantus are so well done the original Greek text can almost be reconstructed from them. In translating Halley, I kept that in mind as a kind of ideal; however, I did not wish to martyr the text to the ideal. So, while I tried, for example, to translate consistently—using the same English for the same Latin—I could not follow this through without straining my own ears' tolerance, not to speak of my readers' ears. Thus, for the most part I translate, say, *nempe* as "namely," but sometimes as "naturally"; *vero* I sometimes translate as "in fact," sometimes as "indeed," and other times I simply leave it untranslated. On the other hand, a word such as "given," *datum*, is so important in the text and in Greek mathematical discourse that it demands stubborn consistency. Thus, while "given" could easily serve as a translation of *propositum,* and certainly would sound better to modern ears, I always translate *propositum* literally as "proposed"—the proposed sum of squares, the proposed length, the proposed difference, etc.

Halley's own inconsistencies are telling in their own way. When they are just a matter of style they remind us that Halley belonged to a time when scholars wrote Latin as naturally as they did their own language, maybe even more so; they were not bound by the rigid formulas, which always give away writers writing in languages not their own. There are other inconsistencies in Halley's Latin, though, that may reflect a certain understanding of his material. One of these has to do with how Halley writes what in Greek would be the phrase "*to tetragōnon apo tēs*...." He sometimes writes this as *quadratum ex +ablative* and sometimes *quadratum + genitive*, as if to emphasize once the word *apo* and once the genitive in the Greek. For example, in problem 10, he has *quadratum e data summa diametrorum*... and then later in the same problem, *quadratum summae*. It is apparent that Halley saw these formulations, which I have translated, respectively, as "the square on the sum of the given diameters" and "the square of the sum," as interchangeable. The Greek phrase unambiguously refers to a geometrical situation where a square is conceived as *built on* a certain line. Thus, while the Greek phrase may be much abbreviated, the word *apo* is always present. Saying the "square of AB" could suggest the arithmetic

M.N. Fried, *Edmond Halley's Reconstruction of the Lost Book of Apollonius's Conics*, Sources and Studies in the History of Mathematics and Physical Sciences, DOI 10.1007/978-1-4614-0146-9_6, operation of squaring AB, that is, AB^2.[71] On this reading, one might conclude that, despite Halley's general sensitivity to the difference between the Greek geometrical thinking and the modern algebraic thinking, his own modern mind sometimes got the better of him. Of course he may also be using what he sees as equivalent formulations for precisely the geometrical situation of a "square on AB"; however, since there are other places in Halley's text containing subtle betrayals of algebraic thinking (see, for example, VIII.13, note 86), one must at least consider the former possibility.

Parentheses occur frequently in Halley's text. Generally, these just contain references to propositions in the *Conics* or other works; but occasionally they also contain clarifications. In the translation these are preserved. My own parenthetical remarks, additions, and clarifications are contained by square parentheses. There are two deviations from this. (1) Each proposition in Halley's text is subtitled "Problem." In the text itself this appears thus: "Propositio I. Probl." It seemed to me clearer for modern readers to write "Proposition I (Problem)." (2) I have included the original Latin word or phrase in rounded parentheses where it seemed to me that the English translation did not catch entirely the nuances of Halley's Latin.

As for the diagrams, except for the extra smoothness and straightness one can achieve with modern technology, the diagrams below were meant to be entirely faithful to those in the 1710 edition. I have not numbered the diagrams since the text, for the most part, makes it clear which diagram is being referred to. That said, Halley sometimes refers to diagrams in a previous proposition by saying, at the start, something like: "Let things in the hyperbola be kept as they were previously, and in the same way as was demonstrated in the preceding propositions...." Yet, there is rarely any difficulty here both because the diagram usually appears in the previous proposition or the one just before it and because Halley is very consistent in how he draws and letters his diagrams. Where there is any ambiguity at all, however, I have indicated which diagram he intends in a footnote.

Since Halley follows Book VII of the *Conics* step-by-step in his reconstruction of Book VIII, as I have described above, readers of the reconstruction will benefit greatly by having a copy of Apollonius's *Conics*, Book VII by their side for reference and comparison. But this is not always possible or convenient. For this reason, in the notes, I have provided full statements of propositions from Book VII that Halley refers to in the text.[72] I grant this may sometimes be more than the reader truly needs; however, I would rather sin on the side of excess than leave a reader frustratingly wondering about a word or formulation that I thought unnecessary to include.

---

[71] Interestingly enough, when Heiberg translates the Greek phrase *to apo+genitive*, he, like Halley, also often uses the formula *quadratum+genitive*; however, tellingly, when it comes to the *ekthesis*, Heiberg will always write "AB^2" where the Greek has *to apo AB*.

[72] As noted above, I will always use Toomer's masterful edition of the *Conics*, Books V–VII (Toomer, 1990) for this purpose.

Part II
Apollonius of Perga's *On Conics*: Book Eight Restored

APOLLONIUS OF PERGA'S *ON CONICS*: BOOK EIGHT RESTORED OR THE BOOK ON DETERMINATE PROBLEMS CONJECTURED

Halley to Aldrich, S. P.[73]

When I had set out to publish the [Conics] of Apollonius, it disturbed us not to a small degree that in the Arabic Codex the last book of the Conics was missing. But you, as it is by [your] happy nature, felt at once that there is no need to deplore [the loss], but that surely to some degree it could be restored, given the fact that in Pappus' Mathematical Collection, [Pappus] himself passed on lemmas serving [what] was to be demonstrated in the seventh and the eighth book of the Conics at the same time; whereas in the other books, different [lemmas] to the different [books] are provided. Hence, to you we owe the discovery that the [two] books are conjoined; that the problems in [Book] eight are allotted their determination from the theorems on diorisms (διοριστικοῖς)[74] *in [Book] seven. Indeed, having weighed the matter carefully, this seemed to me to be indicated not only by probable conjecture but also by certain [concrete] vestiges: with that, and with you showing the way, I undertook, as far as I could, to restore the loss. Therefore, I pray, please accept whatever I have managed to accomplish. Farewell.*

[73] Henry Aldrich (1648–1710) was dean of Christ Church College, Oxford from 1689 until his death in 1710. It was Aldrich who persuaded Halley to collaborate with David Gregory in preparing an edition of the works of Apollonius. See the introduction above and Suttle (1940) for more details about Aldrich. The letters S.P. stand for *salutem plurimam*, "many greetings." It is a standard Latin epistolary formula: I decided to leave this untranslated to emphasize its formulaic character.

[74] This, indeed, is how Apollonius refers to Book VII in the letter introducing Book I. However, *diorisms* or limits of possibility, limits within which a given problem can be solved, are not referred to as the main topic of Book VII in the letter introducing Book VII. There, Apollonius describes the book as follows: "In this book [Book VII] are many wonderful and beautiful things on the topic of diameters and the figures constructed on them, set out in detail. All of this is of great use in many types of problems, and there is much need for it in the kind of problems which occur in conic sections which we mentioned, among those which will be discussed and proven the eighth book of this treatise." (Here, and throughout, translations from Book VII are from Toomer (1990). Also, unless indicated otherwise (i.e., by my initials, MNF), bracketed remarks in quotations from Toomer are Toomer's own.)

M.N. Fried, *Edmond Halley's Reconstruction of the Lost Book of Apollonius's Conics*, Sources and Studies in the History of Mathematics and Physical Sciences,
DOI 10.1007/978-1-4614-0146-9_7,

Proposition I (Problem)[75]

Given the latus rectum of any diameter in a given parabola, exhibit the latus rectum of any other diameter.

Since it has been shown (by VII.5)[76] that the latus rectum of any other diameter of a parabola exceeds the latus rectum of the axis by four times the intercept [along the axis] between the perpendicular dropped to the axis [from the vertex of the other diameter] and the principal vertex of the section [i.e., the vertex of the axis], it is clear that if a point is taken beyond the vertex of the axis at a distance of one-fourth the latus rectum, the portion of the axis between that point and the perpendicular dropped from the vertex of any other diameter will be one-fourth the latus rectum of that diameter.

So let B be the vertex of parabola $AB\Gamma$, $B\Delta$ its axis, and BZ the latus rectum [of the axis]; and let the axis be produced to E, so that EB is made one-fourth the latus rectum of the axis; and through any point Γ of the section let ΓH be drawn parallel to the axis, which, accordingly, will be a diameter (by I.46); also let a perpendicular $\Gamma\Theta$ be dropped to the axis. I say that the latus rectum of diameter ΓH is four times the intercept ΘE.

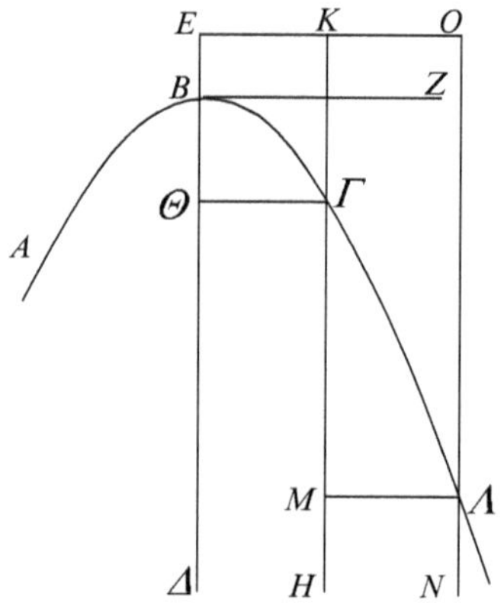

But if the given diameter, as ΓH, is not the axis, let ΓH be produced beyond the vertex to a point K such that ΓK is made one-fourth the latus rectum of the given diameter; from any point Λ on the section let a perpendicular, as ΛM, be dropped to [the diameter]; the latus rectum of the diameter ΛN will be four times the intercept

---

[75] Every proposition in the reconstruction.

[76] *Conics* VII.5 states: "If there is a parabola, and one of its diameters is drawn in it, and from the vertex of that diameter a perpendicular is drawn to the axis, then the parameter of the lines drawn from the section to the diameter parallel to the tangent drawn from the vertex of the diameter, [i.e., when figures are applied to that line, [those ordinates] are equal in square to them—and that line is its [the diameter's] *latus rectum*—is equal to the *latus rectum* of the axis increased by four times the amount cut off from it by the perpendicular from the axis adjacent to the vertex of the section."

KM. All these things are perfectly clear from the fifth proposition of the seventh book of this [work (i.e., the *Conics*)].[77]

Proposition II (Problem)

Conversely, given the latus rectum of any diameter in a parabola, find the diameter whose own latus rectum is equal a given line.

Let the given parabola be $AB\Gamma$,[78] but this time let the latus rectum of a given diameter ΓH be given. Let ΓH be produced to K so that ΓK is made one-fourth the latus rectum of this diameter [i.e., of ΓH]; and let KM be set equal one-fourth the latus rectum of some other diameter; and let a perpendicular through M be set up meeting the section at Λ; and let ΛN be drawn through Λ parallel to ΓH. I say that ΛN, which has been produced to O, is the diameter of the section sought.

Alternatively, if a line EKO is drawn through K perpendicular to the diameter, and to the interval $O\Lambda$, which is equal to one-fourth the given latus rectum, a line $M\Lambda$ is drawn parallel to EKO, it will meet the section at the point sought Λ. For while ΓK is one-fourth the latus rectum of ΓH, also KM, to which ΛO is equal, is one-fourth the given latus rectum; but ΛO (by what was demonstrated in VII.5) is one-fourth of the latus rectum of diameter ΛN: ΛN therefore will be the diameter which we are looking for.

The given latus rectum, however, cannot be less than the latus rectum of the axis (by VII.32).[79]

Proposition III (Problem)

Given the latus rectum of any given diameter in a given hyperbola, find the latus rectum of any other given diameter.

Let ΓB be some given diameter in a hyperbola BE, whose center is A, and let ZB equal to its semi-latus rectum: we are asked to investigate the latus rectum of any other diameter ΔE.

Through the three given points B, E, Z, (by *Elem.* IV.5)[80] let a circle $EBZH$ be described cutting the diameter set forth, ΔE, at point H. I say EH is half the latus rectum sought.[81] For (by VII.29) the differences between the squares of any of the

[77] Let the intersection of BZ and KH be X. Then, by VII.5, lr.(ΛN)=$4XM$+lr.(axis) $= 4XM + 4EB = 4XM + 4XK = 4KM$ where I am using "lr.(ΛN)" as an abbreviation for "the latus rectum of diameter ΛN", "lr.(axis)" for "the latus rectum of the axis," and so on. Where there is no risk of confusion, I shall just write "lr" for the latus rectum in question.

[78] There is no separate diagram for proposition II; Halley is still referring to the diagram for the previous proposition. See the remarks on the diagrams in the introduction.

[79] VII.32 states the *diorism*: "[In] every parabola, the *latus rectum* which is the parameter of the ordinates falling on the axis is the smallest of the *latera recta* which are the parameters of the ordinates falling on the other diameters; and the *latus rectum* which is the parameter of the ordinates falling on [one of] those diameters closer to the axis is less than the *latus rectum* which is the parameter of the ordinates falling on [a diameter] farther [from it]."

[80] *Elements* IV.5: "About a given triangle to circumscribe a circle." Unless noted otherwise, I am taking advantage of Heath's (1926) translations.

[81] Halley's use of "semi-latis rectum" in the opening of the proposition and "half of the latus rectum" here is an example of the sort of inconsistency that I described in the introduction.

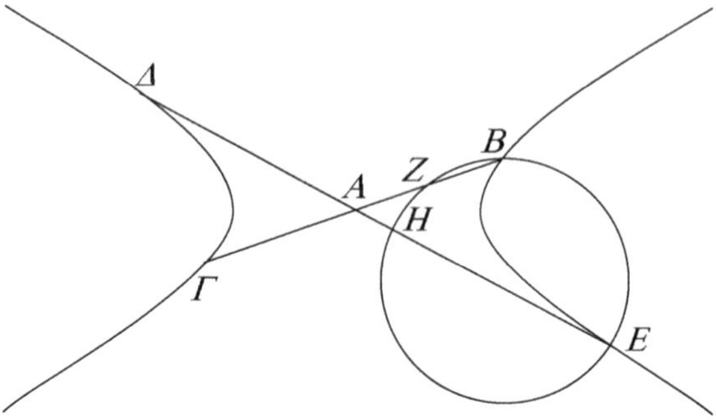

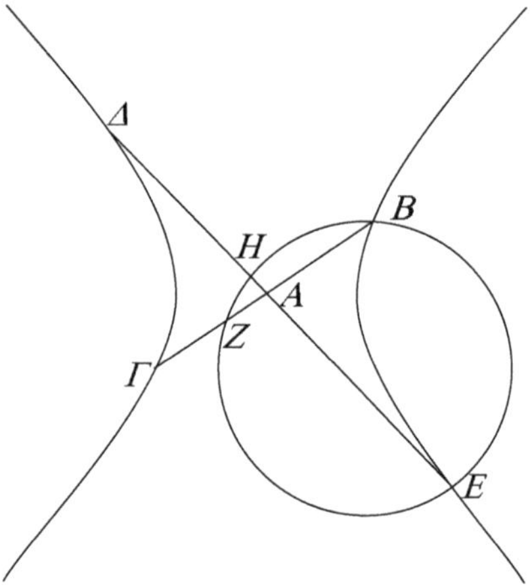

diameters of a hyperbola and the figures on them are everywhere the same[82]: and, with that, the rectangles contained by the diameters and the difference between the same diameters and their latera recta are always equal, so that also the rectangles contained by half of the same [lines] are equal[83]; whence, rectangle *BAZ*[84] is equal to that contained by *AE* and the difference between *AE* and the semi-latus rectum of diameter ΔE. But, on account of the circle (by *Elem.* 35 or 36),[85] the rectangle

I reemphasize here that I try to preserve Halley's own formulations, including such inconsistencies, as far as possible throughout the translation.

[82] This is the enunciation, essentially, of VII.29.

[83] *Conics* VII.29 tells us that if D and D′ are any two diameters of a hyperbola, and L and L′ are their respective *latera recta*, then sq.D-rect.D,L=sq.D′-rect.D′,L′. Hence, rect.D,(D-L)=rect.D′,(D′-L′), from which it also follows that rect.½D,(½D-½L)=rect.½D′,(½D′-½L′).

[84] That is, the rectangle contained by the segments, *BA* and *AZ*. Halley, in this, is following Apollonius's and Euclid's usual notational practices.

[85] *Elements* III.35 (which Halley needs for the lower diagram) states: "If in a circle two straight lines cut one another, the rectangle contained by the segments of the one is equal to the rectangle contained by the segments of the other"; *Elements* III.36 states: "If a point be taken outside a circle and from it there fall on the circle two straight lines, and if one of them cut the circle and the other touch it, the rectangle contained by the whole of the straight line which cuts the circle and

contained *EA*, *AH* is equal to the rectangle *BAZ*. Therefore, *AH* is the difference between the semi-diameter *AE* and the semi-latus rectum of that same diameter, and, accordingly, *EH* is half the latus rectum sought.

Proposition IV (Problem)

If in a given ellipse a diameter is given together with its latus rectum, we can exhibit the latus rectum of any other diameter.

Let *ΓΕΒΔ* be the given ellipse, whose center is *A*; let the latus rectum of diameter *ΓΒ* be given, and let *BZ* be made equal to half of the latus rectum and placed along the extension of the diameter: and let *ΔE* be any other diameter: and through the three points *E*, *B*, *Z* let a circle be described cutting that same diameter extended at point *H*. I say that *EH* is half the latus rectum sought.

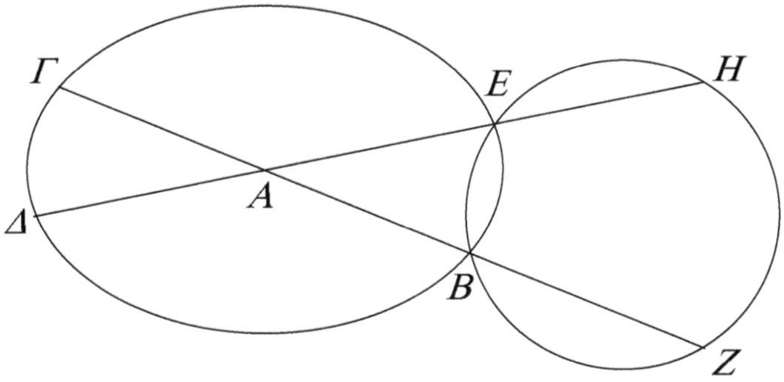

For in an ellipse (by VII.30), the sum of the square of any diameter and the figure on it is always equal,[86] that is, the rectangle contained by the diameter and the diameter and its latus rectum together will everywhere be equal, so that also the rectangles contained by the halves of these same [lines] will [everywhere be equal].[87] The rectangle *BAZ*, therefore, is equal to that contained by *EA* and *EA* and its semi-latus rectum together. But (by *Elem.*, III.36) the rectangle *EAH* is equal to the rectangle *BAZ*; whence, *AH* is equal to the semi-diameter *EA* taken together with its semi-latus rectum; wherefore *AH* exceeds *EA* by half the latus rectum sought; twice the line *EH*, therefore, is equal to that latus rectum.

---

the straight line intercepted on it outside between the point and the convex circumference will be equal to the square on the tangent." It follows immediately (and it is this result, which is not stated explicitly in Euclid, that Halley needs for the upper diagram) that if two straight lines in a circle cut one another *outside* the circle the rectangle formed by the whole of the one straight line which cuts the circle and the segment of it intercepted on it outside between the point of intersection and the convex circumference is equal to the rectangle formed by the same segments of the other straight line.

[86] Again, this is the enunciation of VII.30.

[87] The reasoning is the same as in the previous proposition, namely, *Conics* VII.30 tells us that if D and D′ are any two diameters of an ellipse, and L and L′ are their respective *latera recta*, then sq.D+rect.D,L=sq.D′+rect.D′,L′. Hence, rect.D,(D+L)=rect. D′,(D′+L′), from which it also follows that rect.½D,(½D+½L)=rect.½D′,(½D′+½L′).

Proposition V (Problem)

Given the sides of the figure of the axis[88] *of a hyperbola and given some other diameter in magnitude, determine the position of that diameter in the section and the position and magnitude and also the latus rectum of the diameter conjugate to it.*

Let AB be a hyperbola, $A\Gamma$ the transverse axis, and $A\Delta$ the latus rectum; let the later [i.e., the latus rectum] be placed in the direction of the axis from Δ and δ[89]; and let it be contrived that as ΓN is to NA and $A\Xi$ is to $\Xi\Gamma$ so too ΓA is to $A\Delta$; the points N, Ξ, therefore, will be (by VII.2)[90] the ends of the lines which we have called the *homologues.*

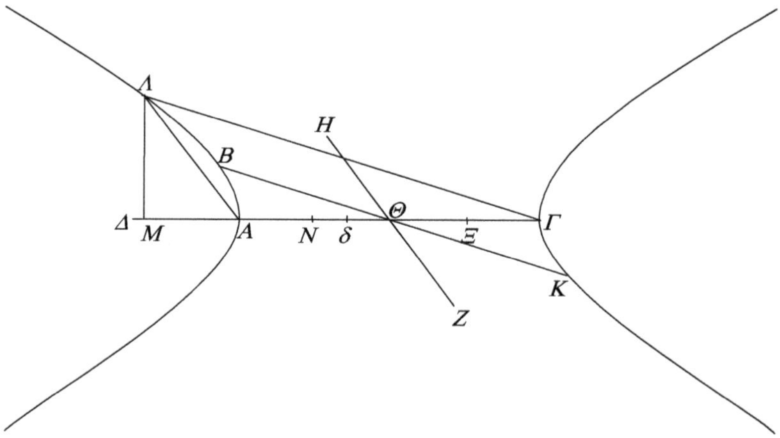

[88] By "the figure of the axis," Halley means the figure *with respect to* the axis, that is, the rectangle whose sides are the axis and the latus rectum with respect to the axis (for this reason the latus rectum is called the latus rectum, the upright side—it is the upright side of that rectangle!).

[89] Halley means by this that a segment equal to the latus rectum is drawn from these points to A, so that, $\Delta A = A\delta$.

[90] Proposition VII.2 is cited only because in the course of it the *homologue* is first introduced. The complete proposition runs as follows: "If the axis in a hyperbola is extended in a straight line so that the part of it falling outside of the section is the transverse diameter, and a line is cut off adjoining one of the ends of the transverse diameter such that the transverse diameter is divided into two parts in the ratio of the transverse diameter to the *latus rectum*, while the line cut off corresponds to the *latus rectum*, and a line is drawn from that end of the transverse diameter which is the end of the line which was cut off to the section, in any position, and, from the place where [that line] meets it [the section], a perpendicular is drawn to the axis, then the ratio of the square on the line drawn from the end of the transverse diameter to the rectangle contained by the two lines between the foot of the perpendicular and the two ends of the line which was cut off is equal to the ratio of the transverse diameter to the excess of it [the transverse diameter] over the line which was cut off.

"And let the line which was cut off be called the 'homologue'."

But, *componendo*,[91] ΓA will be to AN as the transverse axis and the latus rectum together will be to the latus rectum, or as $\Gamma\Delta$ will be to ΔA; indeed, by conversion, AN will be to $N\Xi$ as $A\Delta$ to $\delta\Gamma$,[92] the difference between the axis and the latus rectum. Therefore, *ex aequali*, ΓA will be to $N\Xi$ as $\Gamma\Delta$ is to $\delta\Gamma$, and, accordingly, the rectangle contained by $N\Xi$ and $\Gamma\Delta$ will be equal to that contained by $A\Gamma$ and $\Gamma\delta$ or by the axis and the difference between the axis and its latus rectum. But the rectangle contained by $A\Gamma$ and the difference between it and its latus rectum is equal to the difference between the square of the axis and its figure, which is, in fact, (by VII.13 and 29) the difference between the squares on any conjugate diameters on the section.[93] With that, the rectangle contained by $N\Xi$, $\Gamma\Delta$ is equal to the difference between the squares on any conjugate diameters of the section. Now, suppose BK is the diameter that we are seeking. Repeating the figure from proposition VII.6 (and by the same proposition),[94] the square on BK will be to the square on ZH as ΞM is to MN, and, by conversion, BK will be to the difference of the squares on BK, ZH, or, rather, to the rectangle contained by $N\Xi$ and $\Gamma\Delta$, as ΞM will be to $N\Xi$. For this reason, that containing $N\Xi$ and the square on BK will equal that contained by $N\Xi$ and the rectangle contained by $\Gamma\Delta$ and ΞM; and, with that, the square on BK will

---

[91] It is worthwhile to review at this point some of the conventional words used to describe the manipulations of proportions. Let a:b::c:d. Then the proportion obtained *componendo* is a+b:b::c+d:d; the proportion obtained *separando* is a-b:b::c-d:d; the proportion obtained *convertendo* (Halley typically writes, simply, "by conversion," *per conversionem*) is a:a-b::c:c-d; the proportion obtained, *permutando* or *alternando*, is a:c::b:d (provided a,b,c,d are magnitudes of the same kind). If a:b::c:d and b:e::d:f, then the proportion obtained *ex aequali* is a:e::c:f. These relationships and others are described by Euclid in the definitions opening Book V of the *Elements*.

[92] Remember $AN = \Gamma\Xi$ and $A\Delta = A\delta$. So, with that in mind, we have from $NA:\Gamma N::\Xi\Gamma:EA::A\Delta:A\Gamma$, $NA:N\Xi::\Xi\Gamma:A\Xi - \Xi\Gamma::A\Delta:\Gamma A - A\Delta::A\Delta:\Gamma\delta$.

[93] It is not necessary to apply *both* VII.29 and VII.13. For VII.13 states that, "[In] every hyperbola, the difference between the squares on its axes is equal to the difference between the squares on any pair of its other conjugate diameters whatever." So, if D and d are conjugate diameters, while X and x are the conjugate axes, then VII.13 tells us that sq.D-sq.d=sq.X-sq.x. But, by the definitions following I.16 and proposition I.60, we have sq.x=rect.X,lr.X (where "lr.X" indicates the latus rectum of the figure on X—a notation I shall continue to use for convenience). Therefore, sq.D-sq.d=sq.X-sq.x=sq.X-rect.X,lr.X. Alternatively, sq.D-sq.d=sq.D-rect.D,lr.D by I.60, while sq.D-rect.D,lr.D=sq.X-rect.X,lr.X by VII.29; therefore, again, sq.D-sq.d=sq.X-rect.X,lr.X.

[94] VII.6 states: "If there are constructed on the extension of the axis of a hyperbola two lines adjacent to the two ends of the axis which is the transverse diameter, each of them being equal to the line which we called 'homologue', and placed as it [the homologue] is placed, and two conjugate diameters from among the diameters of the section are drawn, and, from the vertex of the section, a line is drawn parallel to the 'erect' diameter of the two to cut the section, and from the place where it intersects it a perpendicular is drawn to the axis: then the ratio of the transverse of the two conjugate diameters to the erect one is equal in square to the ratio of the line between the foot of the perpendicular and the end of the more remote of the two homologues to the line between the foot of the perpendicular and the end of the nearer of the two homologues; and the ratio of the transverse diameter to the parameter of the lines drawn to it [the transverse diameter] parallel to the second diameter, which [parameter] is its *latus rectum*, is, in length, equal to the ratio of the two lines which we mentioned previously to each other in length." The "specification" (διορισμός) is precisely what Halley goes on to say.

be equal to the rectangle contained by ΞM and $\Gamma\Delta$,[95] or, rather, contained by ΞM and the sum of the axis and its latus rectum. Therefore, BK is the mean proportional between $\Gamma\Delta$ and $M\Xi$. But BK and $\Gamma\Delta$ are given; therefore $M\Xi$ is given, and, the point Ξ being given, the point M is given.

And so, the synthesis is clear. For in keeping with what has been described, let it be contrived that as $\Gamma\Delta$ is to BK so BK is to $M\Xi$, and with ΛM set up perpendicular to the axis, let it be contrived that the square on ΛM be to the rectangle $AM\Gamma$ [96]as the latus rectum of the figure on the axis is to the transverse side, and the point will touch the section (by proposition 21 of Book I).[97] Let ΛA, $\Lambda\Gamma$ be joined, and, through the center Θ, let the diameters BK, ZH be drawn parallel to ΛA, $\Lambda\Gamma$: hence, we have (by what is demonstrated in VII.6) the positions of each of the diameters sought; and ΘB having been made equal to the given semi-diameter, the point B touches the section. But let it be contrived that the square on ΘZ, or ΘH, to the square on ΘB is as MN to $M\Xi$, and ZH will be the diameter conjugate to BK: and (by VII.6) the given diameter BK will be to its latus rectum as $M\Xi$ is to MN.

Therefore, without the curve being described beforehand, we find the position of diameter BK and both the position and magnitude of its conjugate, and its latus rectum: this is to be observed in all the rest. For it appears that Apollonius had resolved to put the seventh book to this use, to pave the way toward the solutions and determinations of all problems which pertain to the sums or differences of conjugate diameters or the sides of their figures, for example, the sums or differences of the squares on these or similar [lines], without assuming a sketch of the section: nor did he pursue the required positions of the diameters by ordinary methods (*vulgari artificio*).

But you will find the magnitude and latus rectum of the conjugate diameter of a given diameter by a somewhat simpler construction, provided you do not require its position. For the mean proportional may be obtained between $A\Gamma$ and $\Gamma\delta$, or, rather, between the axis and the difference between the axis and its latus rectum: that, accordingly, will be equal in square (*poterit*)[98] to the difference between the

---

[95] Despite Halley's general care in maintaining Apollonius's style, this is a manifestly un-Apollonius manipulation, and it is clearly algebraically motivated. He argues as follows: sq.BK:rect.$N\Xi,\Gamma\Delta$::ΞM:$N\Xi$, therefore—and now I shall shift to an algebraic notation (even though Halley wants to say rect.$N\Xi$,(sq.BK)!)—$N\Xi\cdot BK^2=\Xi M\cdot(N\Xi\cdot\Gamma\Delta)=N\Xi\cdot(\Xi M\cdot\Gamma\Delta)$, from which we have $BK^2=\Xi M\cdot\Gamma\Delta$.

[96] As before, rect. $AM\Gamma$ means rect.AM,$M\Gamma$.

[97] In fact, it is the converse of I.21. I.21 states: "If in an hyperbola or ellipse or in the circumference of a circle straight lines are dropped ordinatewise to the diameter, the squares on them will be to the areas contained [τὰ παριεχόμενα χωρία] by the straight lines cut off by them beginning from the ends of the transverse side of the figure, as the upright side of the figure is to the transverse, and to each other as the areas contained by the straight lines cut off (abscissas), as we have said." For quotations from Books I–III, I shall take advantage of Taliaferro's translations (as revised by Dana Densmore in the Green Lion edition).

[98] Halley's use of the verb *posse* here, and in similar passages, is exactly parallel to the use of the Greek *dynasthai*, which, in Greek mathematical works, means "equal in square" (see the remarks in the introduction).

square on the axis and its figure, that is (by VII.29),[99] the difference between the squares on any conjugate diameters. Let it be ΠO, let ΠP be set up perpendicular to $O\Pi$, and let OP be made equal to the given diameter, and ΠP will be equal to the diameter conjugate to PO if, of course, the latus rectum was less than the axis.[100] If it is greater, let ΠP be made equal to the diameter set forth, and when OP has been joined it will equal its conjugate diameter; and with $O\Sigma$ set up perpendicular to OP above, $P\Sigma$ will be the latus rectum of diameter ΠP. In the prior case, if a perpendicular ΠT is dropped, PT will be the latus rectum of diameter OP: for the conjugate is the mean proportional between the diameter and its latus rectum. The rest is clear from VII.13 and 29.

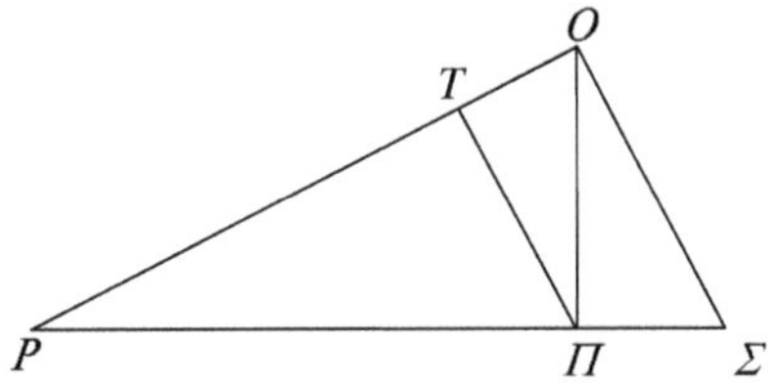

And it is manifest that the diameter set forth ought not to be less than the axis of the hyperbola.[101]

Proposition VI (Problem)

Given the axis of an ellipse and its latus rectum, and given some other diameter in magnitude, describe the position of this diameter and the position and magnitude of the diameter conjugate to the given diameter and also its latus rectum [i.e., of the conjugate diameter].

Let $A\Gamma$ be the transverse axis, extended in both directions, of an ellipse $AB\Gamma$, and let $A\Delta$, $A\delta$ each be made equal the latus rectum, and let the homologue lines (by VII.3)[102] be AN, $\Gamma\Xi$; then it must be understood, of course, that ΓN is to NA and

---

[99] And by I.60 or the definition of the second diameter, as described above.

[100] ΠO is chosen to be the mean proportional between $A\Gamma$ and $A\Gamma$-lr.$A\Gamma$. Therefore, sq.ΠO=rect.$A\Gamma$,($A\Gamma$-lr.$A\Gamma$)=sq.$A\Gamma$-rect.$A\Gamma$,lr.$A\Gamma$. Let D be the given diameter and d the diameter conjugate to it. Then, by VII.29, sq.ΠO=sq.$A\Gamma$-rect.$A\Gamma$,lr.$A\Gamma$=sq.D-rect.D,lr.D. But, by I.60, rect.D,lr.D=sq.d. Therefore, sq.ΠO=sq.D-sq.d or sq.d=sq.D-sq.ΠO. Thus, setting OP equal to the given diameter, D, PR will, by the Pythagorean theorem, be equal to d. As for the *diorism* at the end, it follows from VII.29, and Apollonius states this explicitly in the corollaries following VII.31, that if the axis of a hyperbola is greater than its latus rectum, then every diameter is greater than its latus rectum.

[101] That the axis is the shortest transverse diameter of a hyperbola is not proven explicitly by Apollonius, but it is stated in VII.28 and follows from V.34 (see Toomer, 1990, p. 603, note 107).

[102] As in the last proposition, VII.3 is cited here only because it introduces the homologue for the ellipse. The proposition itself states: "If a line is constructed on the extension of one of the axes of an ellipse, whichever axis it may be, and one of [the line's] ends is one of the ends of the transverse diameter, while the other end is outside of the section, and the ratio of [the line] to the line between its other end and the remaining end of the [transverse] diameter, and a line is drawn from the

$A\Xi$ to $\Xi\Gamma$ as $A\Gamma$ is to $A\Delta$. Hence, *dividendo*,[103] ΓA will be to AN as the difference between the axis and its latus rectum will be to the latus rectum, or as $\Gamma\delta$ to ΔA; *componendo*, AN will be to $N\Xi$ as the latus rectum to the sum of the axis and its latus rectum, or as $A\Delta$ is to $\Delta\Gamma$ [104]: therefore, *ex aequali*, ΓA will be to $N\Xi$ as $\Gamma\delta$ to $\Delta\Gamma$, or as the difference between the axis and its latus rectum to the sum of the same.

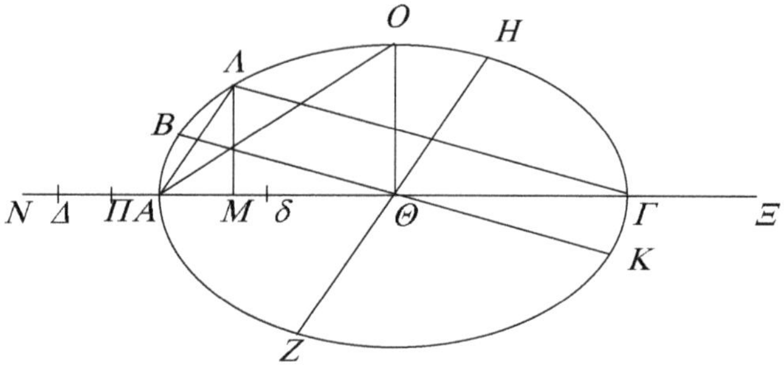

Therefore, the rectangle contained by the axis and the sum of the axis and its latus rectum is equal to the rectangle contained by $N\Xi$ and the same difference [i.e., the difference between the axis and the latus rectum]. But the rectangle contained by the axis and the axis and latus rectum together is equal to the square on the axis and its figure together; in fact, this sum is equal to the sum of the squares on any conjugate diameters of the section (by VII.30 and 12).[105] Therefore, the rectangle contained by the given [lines] $N\Xi$, $\Gamma\delta$ or the difference of the axis and the latus rectum.

Consider now the task at hand (*factum quod quaeritur*), and let BK be equal to the given diameter and ZH be conjugate to it; let $\Lambda\Gamma$ be drawn parallel to [BK], let ΛA be joined and perpendicular ΛM be dropped: and so (by what was demonstrated

---

end common to the [transverse] diameter and the line constructed on the axis, to any point on the section, and from the place where it meets [the section] a perpendicular is drawn to the axis, then the ratio of the square on the line which was drawn [to the section] to the rectangle contained by the two lines between the foot of the perpendicular and the two ends of the [first] line which was constructed on the axis is equal to the ratio of the transverse diameter to the line between those two ends of the transverse diameter and the line which was constructed that are different from each other. Let the line which was constructed be called the 'homologue'."

[103] This is the same as *separando*, defined above.

[104] By definition of the homologue, $\Xi\Gamma{:}A\Xi :: A\Delta : A\Gamma ::$ lr.(Axis) : Axis. So *componendo*, $\Xi\Gamma$: $\Xi\Gamma + A\Xi{::}A\Delta{:}\Delta A + A\Gamma{::}$lr.(Axis): Axis + lr.(Axis).

[105] Proposition VII.12 is the proposition for the ellipse corresponding to VII.13, which we have seen, for the hyperbola: "[In] any ellipse, the sum of the squares on any two of its conjugate diameters whatever is equal to the sum of the squares on its two axes." So, if D and d are conjugate diameters and X and x are the conjugate axes, VII.12 tells us that sq.D+sq.d=sq.X+sq.x. But, by I.15, sq.x=rect.X,lr.X; therefore, sq.D+sq.d=sq.X+rect.X,lr.X. Alternatively, by I.15, sq.D+sq.d=sq.D+rect.D,lr.D. But by VII.30 sq.D+rect.D,lr.D=sq.X+rect.X,lr.X. So, again VII.12 or VII.30 is needed, but not both.

in VII.7)[106] the square on BK will be to the square on ZH as ΞM to MN. But, *componendo*, the square on BK will be to the sum of the squares on BK and ZH, that is, to the rectangle contained by $N\Xi$ and $\Gamma\delta$, or, rather, the difference between the axis and its latus rectum, as ΞM is to $N\Xi$. Therefore, that contained by $N\Xi$ and the square on BK will be equal to [that contained] (*facto*) by $N\Xi$ and the rectangle contained by $M\Xi$ and $\Gamma\delta$; whence the square on BK is equal to the rectangle contained by $M\Xi$ and $\Gamma\delta$, and, accordingly, BK is the mean proportional between $M\Xi$ and $\Gamma\delta$. But BK and $\Gamma\delta$ are each given: therefore, $M\Xi$ is given; and because the point Ξ is given, the point M is given as well.[107]

And so, the synthesis will be accomplished in this way. Let it be contrived that $\Gamma\delta$, or, rather, the difference between the axis and its latus rectum, will be to the given diameter BK as the same BK is to ΞM : and, with $M\Lambda$ set up perpendicularly, let it be contrived that square on $M\Lambda$ will be to the rectangle $AM\Gamma$ as the latus rectum of the axis is to the axis itself, and the point Λ will touch the section (by proposition 21 of Book I). Let $\Lambda\Gamma$, ΛA be joined, and, through the center Θ, let diameters $B\Theta K$, $Z\Theta H$ be drawn parallel these very [lines][108]; and (by what was demonstrated in VII.7), the sought diameters in position are obtained (*habebuntur*); and, with that done, ΘB and ΘK are each equal to the given semi-diameter, so that the points B, K touch the section. By the same VII.7, however, $M\Xi$ will be to MN as BK to its latus rectum, and the square on BK to the square on the diameter ZH conjugate to BK. This problem, therefore, is satisfied, and the position of each of the diameters has been found, and the ellipse was not described beforehand.[109]

But without having to provide the position, find the magnitude of the conjugate diameter and latus rectum. Since the squares on BK, ZH [taken together] are always equal to the squares on the axes taken together, that is, to the sum of the square of the axis and its figure (by VII.12 and 30), if $\Gamma\Pi$ is taken as a mean proportional between $A\Gamma$ and $\Gamma\Delta$, $\Gamma\Pi$ will be equal in square (*poterit*) to the sum of the squares BK, ZH or four times the sum of the squares on the semi-axes $A\Theta$, ΘO, that is, the square on AO. Wherefore, if a semi-circle $A\Delta B$ is described with the diameter $\Gamma\Pi$ or radius AO, in which a line AB, equal to the given diameter BK, is inscribed, and $B\Delta$

106 Again, this is the proposition for the ellipse analogous to VII.6 for the hyperbola. VII.7, in full, states: "If there are constructed on the extension of the axis of an ellipse two lines at the two ends of [the axis], each of them being equal to the homologue, and two conjugate diameters are drawn in the section, and, from the vertex of the section, a line is drawn parallel to one of the conjugate diameters so as to meet the section [again], and, from the place where it meets [the section], a perpendicular is drawn to the axis: then the ratio of the diameter which is not parallel to the line which was drawn, to the other diameter, is equal in square to the ratio to each other of the two parts (of the line between the ends of the two homologues which are not the ends of the diameter) into which it is cut by the perpendicular: according to how the two homologues are placed: if [they are found] on the major axis, they are outside the section, and if on the minor axis, then they are on the axis itself.

"And the ratio of the diameter mentioned to the parameter of the ordinates falling on it (which are parallel to the other [conjugate] diameter) is [also] equal to the ratio mentioned."

107 The quasi-algebraic reasoning here is similar to that at the end of the analysis in proposition V.

108 That is, to $\Lambda\Gamma$ and ΛA, respectively.

109 See problem V above.

is joined, $B\Delta$ will equal the conjugate ZH[110]; and with $\Gamma\Delta$ set up perpendicularly above $A\Delta$ until it meets, at Γ, AB produced, $B\Gamma$ will be a third proportional of $AB, B\Delta$, that is, of BK, ZH.[111] Accordingly, $B\Gamma$ will be equal to the latus rectum sought.

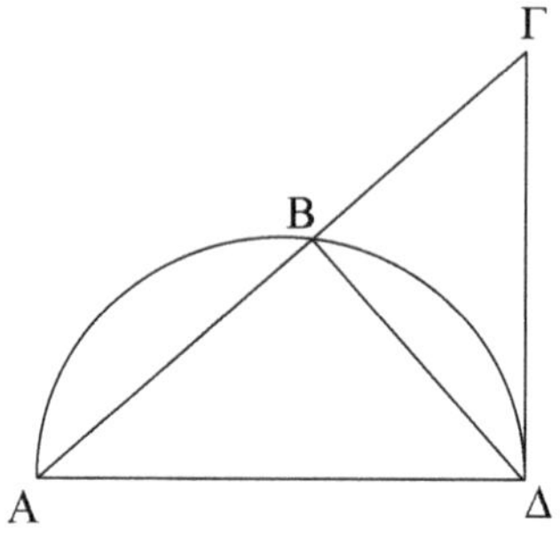

And it is manifest that diameter BK is not greater than the major axis nor less than the minor axis, or the mean proportional between the latus rectum of an axis and that same axis.[112]

Proposition VII (Problem)

Given the axis and the latus rectum of the axis of a hyperbola, and given the ratio of conjugate diameters of the section, find the conjugate diameters both in magnitude and in position.

Let Ξ, N be understood to be the endpoints of the homologues, as was shown in the fifth [proposition] of this [the present work]: and since it can be contrived that the axis be either greater than its latus rectum, or equal, or less, let it first be greater than it, and so also (by what is demonstrated in VII.21)[113] the ratio of the axis to its conjugate ought to be greater than the ratio of any other diameter to its conjugate. Therefore, let the given ratio be as $P\Sigma$ to ΣT, the greater to the lesser, but let it be less than the ratio of the axes among themselves; and let it be contrived that as $P\Sigma$ is to ΣT so too is ΣT to ΣY. Therefore, since $P\Sigma$, ΣT are given, ΣY is also given, and the ratio $P\Sigma$ to ΣY is given as well; indeed, the ratio of the squares on $P\Sigma$, ΣT is

---

[110] Since, by definition, sq.$\Pi\Gamma$=sq.BK+sq.ZH, by construction, $A\Delta=\Pi\Gamma$, $AB = BK$, and because trgl.$AB\Delta$ is right.

[111] In other words, $AB{:}B\Delta{::}B\Delta{:}B\Gamma$ or sq.$B\Delta$=rect.$AB,B\Gamma$, that is, sq.ZH=rect.$BK,B\Gamma$.

[112] If BK were the mean proportional between the major axis, say, and its latus rectum, then the problem would be reduced merely to finding the minor axis.

[113] Proposition VII.21 states: "If there is a hyperbola, and its transverse axis is greater than its erect axis, then the transverse diameter of each pair of conjugate diameters among its other diameters is greater than the erect diameter of [that pair]; and the ratio of the greater axis to the lesser axis is greater than the ratio of the transverse diameter to the erect diameter among the other conjugate diameters; and the ratio of a transverse diameter nearer to the greater axis to the erect diameter conjugate with it is greater than the ratio of a transverse diameter farther [from that axis] to the erect diameter conjugate with it."

the ratio of the squares of the diameters that we are seeking.[114] Now, consider [the problem] done, and let BK be the diameter sought and AH its conjugate, and let $\Gamma\Lambda$ be drawn parallel to this BK and let perpendicular ΛM be dropped. Therefore, (by VII.6) the square on BK to the square on ZH will be as $M\Xi$ to MN, and, with that, $M\Xi$ will be to MN as $P\Sigma$ is to ΣY: and, having contrived $\Sigma\phi$ to be equal to $P\Sigma$, *dividendo*, ΞN will be to $M\Xi$ is as ϕY to ΣP. But the ratio ϕY to ΣP is given,[115] and so the ratio ΞN to $M\Xi$ is given: and $N\Xi$ is given, whence $M\Xi$ is given; and for this reason, since point Ξ is given, the point M is also given. But (by what was demonstrated in the fifth [proposition] of this [work]) the diameter BK is the mean proportional between the given [lines] $M\Xi$ and the sum of the axis and its latus rectum; and, with that, BK is given in magnitude. But the ratio BK to ZH is given; wherefore, since BK is given, ZH is also given.

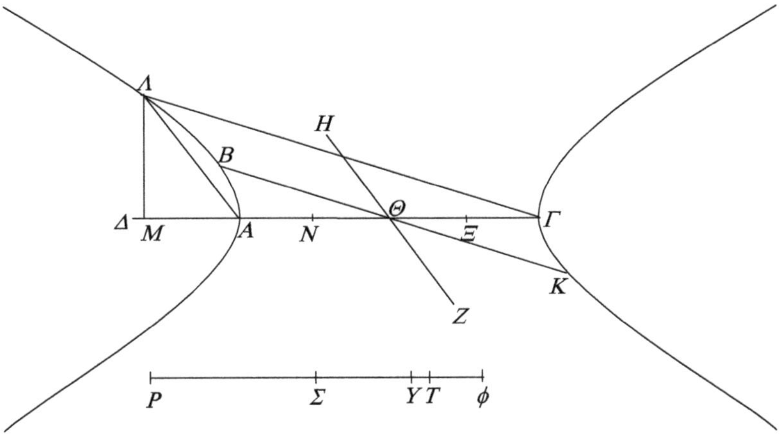

The synthesis of the problem will be accomplished (*componetur itaque problema*), therefore, if it is contrived that ϕY to ΣP, or, rather, the difference between the squares on the terms of the ratio ΣP, ΣT, to the square on the term corresponding to BK,[116] as ΞN is to $M\Xi$; let the perpendicular $M\Lambda$ be set up, let it be contrived that the square on $M\Lambda$ be to the rectangle $AM\Gamma$ as the latus rectum of the axis is to that axis. Let $A\Gamma$, ΛA be joined, and, through the center Θ, let parallels to them, $B\Theta K$, $Z\Theta H$, be drawn: I say that these are the diameters sought, in position, as is clear from the analysis. From here, let BK be taken as the mean proportional

[114] It should be kept in mind that since $P\Sigma : \Sigma T :: \Sigma T : \Sigma Y$, we have $P\Sigma : \Sigma Y ::$ sq.$P\Sigma$:sq.ΣT. Thus, the fact that $P\Sigma : \Sigma Y$ is given means that the ratio of the squares on the diameters is given.

[115] Because $P\Sigma = \Sigma\phi$, ΣY are given, also $\Sigma\phi - \Sigma Y = \phi Y$ and ΣP are given; therefore, the ratio $\phi Y : \Sigma P$ is given.

[116] In other words, sq.$P\Sigma$-sq.ΣT:sq.$P\Sigma$; in referring to $P\Sigma$ as, literally, "the line analogous to BK," I understand Halley to be referring to the positions of $P\Sigma$ and BK in the proportion $P\Sigma : \Sigma T :: BK : ZH$.

between $\Gamma\Delta$ and $M\Xi$, and let half of it be placed from Θ to B and K; these [points] will, accordingly, touch the section; and let it be contrived that as $P\Sigma$ is to ΣT so too will ΘB be to ΘH or ΘZ: and the line ZH will be the diameter conjugate to BK.

Since the difference between the squares on the conjugate diameters is always equal to the difference between the squares on the axes, though, we can give this solution to the problem in a fairly expedient way, but without the position of the diameters. Let the arc of a circle with radius ΣP be described; and, having drawn ΣP from the center, let it be contrived that ΣT be equal to the lesser term of the ratio; let TX be set up perpendicular [to ΣT] cutting the circle at X; let ΣX be drawn, extended if need be, through X. Next, let $T\psi$ be taken, which will be equal in square (*poterit*) to the difference of the squares of the axes of the section, and, at the distance $T\psi$, let $\psi\phi$ be drawn parallel to ΣT, cutting ΣX at ϕ, and let ϕY be drawn parallel to TX: I say $\Sigma\phi$, ΣY are the diameters sought.

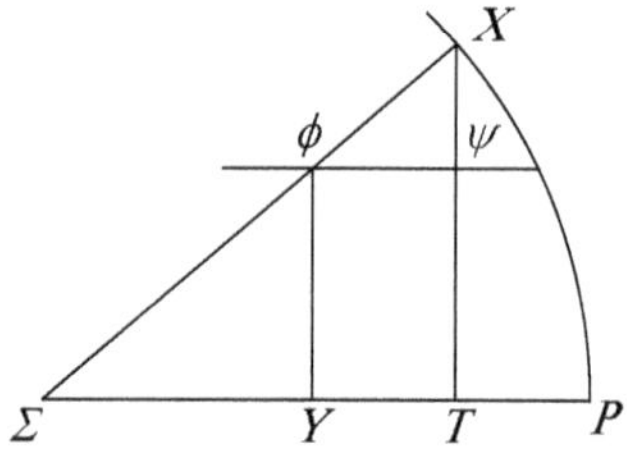

Nor is any other demonstration needed except that ΣY is to $\Sigma\phi$ as ΣT to ΣX, that is, to ΣP, or in the ratio set forth[117]; and that ϕY is in square (*possit*) the difference of squares [formed] from the axes, or the rectangle contained by the axis and the difference between the axis and its latus rectum. And this should be the case by the 13th and equally by the 29th propositions of Book VII.

But if the axis is less than its latus rectum, every thing will be exactly the same both with respect to the analysis and the synthesis: it will be necessary, however, that the ratio set forth be greater than the ratio of the axis to its conjugate axis, though less than the ratio of equal to equal, just as is shown in VII.22.[118] On the other hand, if the axes were equal, all the conjugate diameters will each be equal to one another (by VII.23);[119] and, accordingly, in the so-called equilateral hyperbola, nothing else is learned from the ratio of the conjugates except that it is of equality (*aequalitatis*).[120]

---

117 Which is obviously true.

118 Proposition VII.22 states: "If there is a hyperbola, and its transverse axis is shorter than its erect axis, then the transverse diameter of each pair of diameters among the other conjugate diameters is shorter than the erect diameters of [that pair]; and the ratio of the shorter axis to the longer axis is less than the ratio of any of the other transverse diameters to the erect diameter conjugate with it; and the ratio of a transverse diameter nearer to the shorter axis to the erect diameter conjugate with it is less than the ratio of [a transverse diameter] farther [from that axis] to the diameter conjugate with it."

119 Halley's statement is essentially that of VII.23, namely, "If the two axes of a hyperbola are equal, then every conjugate pair of its diameters is equal."

120 That is, in the ratio that an equal has to an equal.

As for the all the rest, the diameters will approach a ratio of equality, the closer they are to the asymptotes.

Proposition VIII (Problem)

Equally in the case of the ellipse, given the axis and its latus rectum, it is required to find, both in magnitude and position, conjugate diameters having a given ratio to one another.

With the description being kept as it was in Proposition VI, let the given ratio be as $P\Sigma$ to ΣT, and let the diameter to be found, BK, have that ratio, $P\Sigma$ to ΣT, to its conjugate, ZH. Let it be contrived that as $P\Sigma$ is to ΣT so too will ΣT be to ΣY, and $P\Sigma$ will be to ΣY as the square on BK to the square on ZH. In fact, seeing that (by VII.7) the square on BK is to the square on ZH as $M\Xi$ is to MN, $P\Sigma$ will be to ΣY as $M\Xi$ is to MN as well, and *componendo*, YP will be to $P\Sigma$ as $N\Xi$ is to ΞM. But the ratio YP to $P\Sigma$ is given, and, accordingly, the ratio $N\Xi$ to ΞM is given; and, on account of the given $N\Xi$, ΞM is also given, and the point M is given.

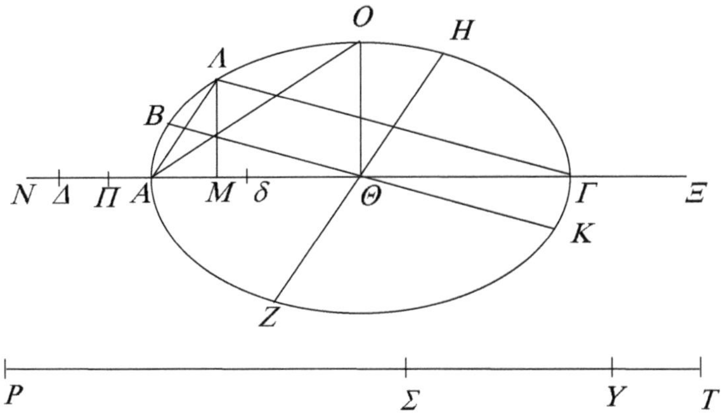

The synthesis of the problem will be accomplished, therefore, if it be contrived that as $P\Sigma$ is to ΣT so ΣT will be to ΣY; next, that it be contrived that as PY is to ΣP, so $N\Xi$ is to ΞM; and, with $M\Lambda$ set up perpendicular [to $A\Gamma$], let [$M\Lambda$'s] square be to the rectangle $AM\Gamma$ as the latus rectum is to the axis $A\Gamma$, let $A\Gamma$, ΛA be joined, which, from what has already been said,[121] will be parallel to the sought diameters, BK, ZH drawn through the center Θ. Therefore, we find each of the diameters in position. Moreover, the mean proportional between $\Gamma\delta$ (the difference between the axis and its latus rectum) and $M\Xi$, just found, will be equal (by proposition VI of this book) to the diameter sought, BK; and if ZH is taken to be to BK in the ratio set forth, or as ΣT is to $P\Sigma$, ZH will be the diameter conjugate to BK; and BK will be to its latus rectum as $P\Sigma$ is to ΣY or as $M\Xi$ to MN, as is plain from VII.7. And these things are to be observed from here on (*in sequentibus*) and there will be no need to repeat them.

[121] As in the synthesis in proposition VI.

But, by what has been proven in the 24th proposition of Book VII,[122] it will be required that the ratio set forth is not greater than the ratio of the major to the minor axis, nor less than the ratio of the minor to the major axis.

But without the position, you will find the magnitude of the diameters in this way. Since their ratio is given, and the sum of the squares on them is equal to the square on the axis together with its figure, that is (by VII.12 and 30), to the sum of the squares of the axes, or four times the square on AO,[123] if the terms of the ratio $P\Sigma$, ΣT are placed in a right angle, and, along [in] the joined [line] PT, $P\phi$ is taken equal to AO, and with center ϕ and radius ϕP, a semi-circle is drawn cutting $P\Sigma$ at point X and diameter PT at point ψ, and $X\psi$ is joined, I say that lines PX, $X\psi$ are equal to the diameters sought.

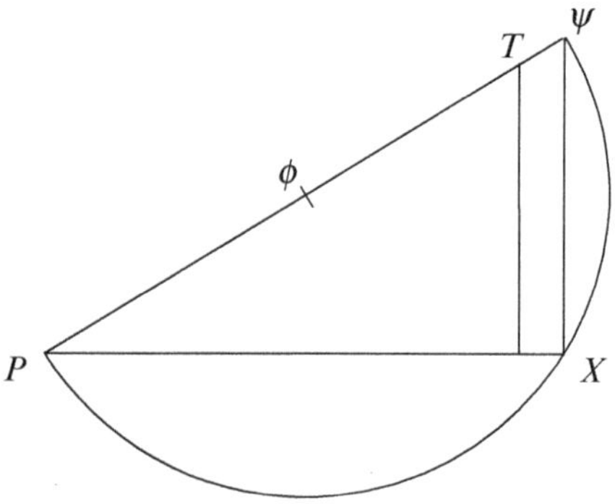

For together, on account of the right angle, they are equal in square (*possunt*) to four times the square on AO, and they are in the ratio of $P\Sigma$ to ΣT; and, accordingly, they are equal to BK, ZH which we are seeking.

Proposition IX (Problem)

Given the axis and latus rectum of a hyperbola, find conjugate diameters of it, both in position and magnitude, whose sum equals a given line.

Repeat the figure in the fifth proposition of this [work], and let points Ξ, N be understood to be the ends of the homologues. Since NA is to ΓN as the latus rectum is to the transverse axis, *componendo*, $A\Gamma$ will be to ΓN as the transverse axis and

[122] Proposition VII.24 states: "If there is an ellipse, and conjugate diameters are drawn in it, then the ratio of the greater of each pair of conjugate diameters to the lesser is less than the ratio of the major axis to the minor axis: and, for any two pairs of conjugate diameters, the ratio of the greater diameter which is nearer to the major axis [than the other greater diameter] to the lesser [diameter] conjugate with it is greater than the ratio of the greater diameter which is farther from the major axis to the lesser diameter conjugate with it." The second condition stated by Halley, namely, that the ratio of the conjugate diameters be not less than the ratio of the minor to major axis, follows immediately from the second part of VII.24.

[123] Since AO is the hypotenuse of the right triangle whose legs are the half-axes $A\Theta$ and ΘO.

the latus rectum together is to the axis itself, $A\Gamma$; and, accordingly, the square on $A\Gamma$ will equal the rectangle $N\Gamma\Delta$, or that contained by $N\Gamma$ and that equal to the axis and latus rectum taken together.

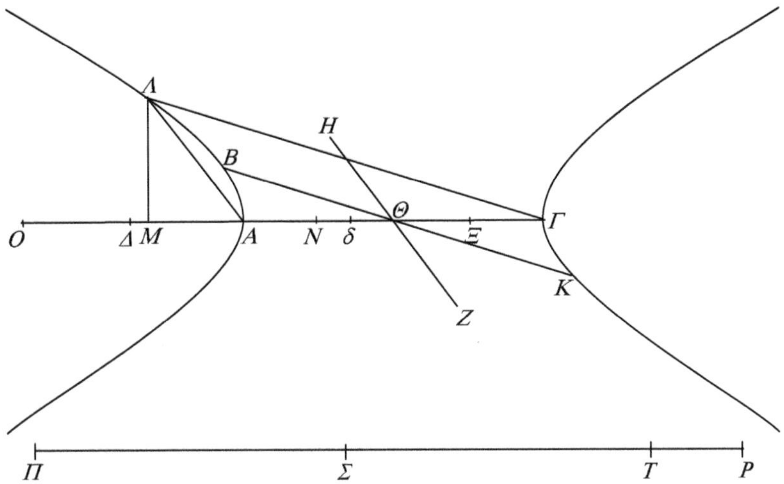

Now, (by VII.8)[124] the square on $A\Gamma$, that is, the rectangle contained by $N\Gamma$, $\Gamma\Delta$ is to the rectangle contained by $N\Gamma$, $M\Xi$ as the square of the sum of the conjugate diameters to the square on $M\Xi$ and that [line] equal in square (*potest*) to the rectangle $NM\Xi$ taken together. Therefore, $\Gamma\Delta$, or the sum of the axis and its latus rectum, will be to $M\Xi$ as the square of the sum of the conjugate diameters to the square on $M\Xi$ and that [line] equal in square (*potest*) to the rectangle $NM\Xi$ taken together, that is (by *Elements* II.4) to the square on $M\Xi$ together with the rectangle $NM\Xi$ and twice the rectangle contained by $M\Xi$ and that equal in square to $NM\Xi$. Therefore, on account of the [line] $M\Xi$, found on both sides, as $\Gamma\Delta$ (the sum of the axis and its latus rectum) is to ΠP (the given sum of the conjugate diameters), so too will the sum of the conjugates be to ΞM, MN taken as one (*simul*) (that is, to twice ΘM) together with twice the [line] equal in square to rectangle $NM\Xi$, and, accordingly, as the half-sum of the conjugate [diameters] to ΘM together with the [line] equal

124 Proposition VII.8 refers to the diagrams in VII.6, 7, which are lettered in precisely the same way as in propositions V and VI in Halley's text, and, with that, it states that "...the ratio of the square on $A\Gamma$ (which is the transverse diameter) to the square on BK and ZH, the two conjugate [diameters], when they are joined together in a straight line, is equal to the ratio of $(N\Gamma \cdot M\Xi)$ [Toomer's notation] to the square on a line equal to line $M\Xi$ plus the line which is equal in square to $(MN \cdot M\Xi)$ [again, Toomer's notation]." Thus, if P is the mean proportional between MN and $M\Xi$, that is, if sq.P=rect.$MN, M\Xi$, then sq.$A\Gamma$:sq.$(BK+ZH)$::rect. $N\Gamma, M\Xi$:sq.$(M\Xi + P)$.

in square to $N M \Xi$.[125] Let that[126] be ΘO, which will be given, because $\Gamma\Delta$ and the sum of the conjugates is given; and, because Θ is given, also the point O is given. Let the common [term] ΘM be removed from both sides[127]: OM [which remains] will equal that [line] equal in square to the rectangle $NM\Xi$, for which reason, MN will be to OM as OM is to $M\Xi$, and, *componendo*, NO will be to OM as $O\Xi$ is to ΞM. But, *permutando*, NO will be to $O\Xi$ as OM to $M\Xi$, from which, again, *componendo*, NO and $O\Xi$ together, or twice ΘO, will be to $O\Xi$ as $O\Xi$ is to ΞM. But since the points Ξ, N, O are given, also the lines ΞO, ON are given; and with that also ΞM is given, and since the point Ξ is given, the point M is given as well.

The synthesis will be accomplished in this way. Having found the points N, Ξ, let it be contrived that as the half-sum of the axis and its latus rectum is to the half-sum of the conjugate diameters so will the same half-sum of the conjugate diameters be to ΘO, which is placed in the extended axis of the hyperbola from the center Θ: next, let it be contrived that twice ΘO to $O\Xi$ is as $O\Xi$ to ΞM; and with the point M already found, set up perpendicular MK, and let the rest be done exactly as was shown in the sixth and seventh propositions above (*praecedentibus*).

But when the points Θ, N, Ξ merge, so that the hyperbola will be equilateral,[128] ΘO and ΞO will be equal; and, accordingly, if it be contrived that as the axis to the given half-sum of the conjugate diameters so too will be the same half-sum to ΘO: and if ΘM is taken as half of ΘO, the point M will be the one we are seeking. This is manifest because of [what has been said] elsewhere, certainly from the fifth [proposition] of this [work].

This problem can be solved though in another way, and not awkwardly. For indeed since the difference of the squares on any conjugate diameters (by VII.13 and 29) be equal to rectangle contained by the axis and the difference between the axis and its latus rectum, and the rectangle contained by the sum of and the difference

---

125 The argument is as follows. With P defined as in the last note, sq.$A\Gamma$: rect. $N\Gamma, M\Xi$:: sq.$(BK + ZH)$:sq.$(M\Xi + P)$. But sq.$A\Gamma$=rect.$N\Gamma, \Gamma\Delta$. Therefore, sq.$A\Gamma$: rect. $N\Gamma, M\Xi$:: rect.$N\Gamma, \Gamma\Delta$: rect. $N\Gamma, M\Xi$, from which we have (by *Elements*. II.4 and the definition of P), $\Gamma\Delta$:$M\Xi$::sq.$(BK+ZH)$:sq.$(M\Xi + P)$:: sq.$(BK+ZH)$:sq.$M\Xi$+sq.P+2rect.$M\Xi, P$::sq.$(BK + ZH)$:sq.$M\Xi$+rect.$M\Xi, MN$+2rect.$M\Xi, P$. So, $\Gamma\Delta$:$M\Xi$::sq.$(BK + ZH)$:sq.$M\Xi$ + rect.$M\Xi$, MN+ 2rect.$M\Xi, P$ or $\Gamma\Delta$:$M\Xi$::rect.$\Pi P, (BK + ZH)$:rect.$M\Xi(M\Xi + MN + 2P)$. The next step, I suspect, was worked out by Halley with the help of algebra (we have seen evidence for that before in the context of proposition V), for by saying $M\Xi$ is "found on both sides" Halley seems to justify the step by "canceling" the common term; however, it *can* be given a classical justification: $M\Xi$:ΠP::rect.$M\Xi(M\Xi + MN + 2P)$: rect.$\Pi P(M\Xi + MN + 2P)$; therefore, since $\Gamma\Delta$:$M\Xi$::rect.$\Pi P, (BK + ZH)$:rect.$M\Xi(M\Xi + MN + 2P)$, we have, *ex aequali*, $\Gamma\Delta$:ΠP:: rect.$\Pi P, (BK + ZH)$:rect.$\Pi P, (M\Xi + MN + P)$ :: $BK + ZH$: $(M\Xi + MN + 2P)$, or $\Gamma\Delta : \Pi P :: BK + ZH : (M\Xi + MN + 2P)$. But $M\Xi + MN = MN + N\Theta + \Theta\Xi + MN$, and $N\Theta = \Theta\Xi$, so that $M\Xi + MN = 2(MN + N\Theta) = 2M\Theta$. Therefore, $\Gamma\Delta : \Pi P :: BK + ZH : 2(M\Theta + P) :: (BK + ZH) : (M\Theta + P)$.

126 Namely, the line $M\Theta + P$.

127 Halley has in mind the equality $\Theta O = OM + M\Theta = M\Theta + P$; thus, removing $M\Theta$ from both side, we obtain, $OM = P$.

128 For in this case the two axes and their latera recta are all equal to one another, so the homologues are equal to the semi-axes.

between conjugate [diameters] be equal (by *Elements* II.6) to the difference between the squares on the same,[129] the rectangle contained by the sum of and the difference between conjugate [diameters] will be equal to the rectangle contained by the axis and the difference between the axis and its latus rectum. Hence, the sum of the conjugate [diameters] will be to the axis as the difference between the axis and its latus rectum is to the difference between the conjugate [diameters]. But everything else being given, the difference between the conjugate [diameters] are given, and, accordingly, the conjugate [diameters] themselves [are given], since both the sum and difference of those are given.

Therefore, let it be contrived that as the proposed sum, ΠP, is to the axis $A\Gamma$, so too will $\Gamma\delta$ be to a fourth proportional, which we will let be PT; and, with ΠT divided in two at Σ, $P\Sigma$ will be the greater of the diameters, $\Pi\Sigma$ the lesser.

But having found the diameters we will obtain their position (by the fifth proposition of this [work]) by taking $M\Xi$ as a third proportional to $\Gamma\Delta$, BK, or NM as a third proportional to $\Gamma\Delta$, ZH.[130] But this (I might add) is the reverse of the synthesis[131]; nor did it occur to *Apollonius* (*nec ad mentem Apollonii*), who, in every single problem before, seems to have sought the point M, from which the rest followed.

But it will be required, by what is demonstrated in proposition 25 of Book VII,[132] that the proposed sum is not less than the sum of the conjugate axes.

Proposition X (Problem)

Given the axis and latus rectum of an ellipse, find the conjugate diameters of it, both in position and magnitude, whose sum is equal to a given line.

As usual (*nempe*), let the ends of the homologue be understood to be N, Ξ; since AN is to $N\Gamma$ as the latus rectum is to the transverse axis, *dividendo*, $A\Gamma$ will be to ΓN as the difference between the axis and latus rectum is to the axis $A\Gamma$; and with that the square of the axis will be equal to the rectangle contained by ΓN and $\Gamma\delta$, the difference between the axis and latus rectum. But (by VII.8) the difference between the axis and the latus rectum, or $\Gamma\delta$, to $M\Xi$ is as the square on the given

---

129 *Elements* II.6 states: "If a straight line be bisected and a straight line be added to it in a straight line, the rectangle contained by the whole with the added straight line and the added straight line together with the square on the half is equal to the square on the straight line made up of the half and the added straight line." So if the half is equal to diameter ZH and the half and the added line is equal to BK, the conjugate to ZH, then the whole line is $BK+ZH$ and the added line is $BK-ZH$. *Elements* II.6 tells us, then, that the rectangle contained by $BK+ZH$ and $BK-ZH$, together with the square on the half, that is, on ZH equals the square on the square on the half and the added line, that is, on BK. Hence, rect.$(BK+ZH),(BK-ZH)+$sq.$ZH=$ sq.BK, or rect.$(BK+ZH),(BK-ZH)+$sq.$ZH=$ sq.$BK-$ sq.ZH.

130 That is, so that $\Gamma\Delta : BK :: BK : M\Xi$ or $\Gamma\Delta : ZH :: ZH : NM$.

131 For above, and in the previous problems, M is found first, and then, having found it, BK and ZH are found.

132 Proposition VII.25 states: "[In] every hyperbola, the line equal to [the sum of] its two axes is less than the line equal to [the sum of] any other pair whatever of its conjugate diameters; and the line equal to the sum of a transverse diameter closer to the greater axis plus its conjugate diameter is less than the line equal to the sum of a transverse diameter farther from the greater axis plus its conjugate diameter."

sum of the conjugate diameters of the ellipse to the square on $M\Xi$ taken together with that [line] equal in square to the rectangle $NM\Xi$.[133] But (by *Elements* II.4) this square is composed (*conficitur*) of the square on $M\Xi$ and the rectangle $NM\Xi$ together with twice the rectangle contained by $M\Xi$ and that [line] equal in square to the rectangle $NM\Xi$. But because $M\Xi$ is found on both sides, as $\Gamma\delta$ is to the sum of the conjugate diameters so too will that same sum be to $M\Xi$, MN taken as one (that is, $N\Xi$) together with twice that [line] equal in square to the rectangle $NM\Xi$; and so too the half-sum of the [conjugate] diameters to $\Theta\Xi$ together with the [line] equal in square to the rectangle $NM\Xi\Theta$. Therefore, $\Theta\Xi$ together with that [line] equal in square to the rectangle $NM\Xi$ is given. Let [the sum] be ΘK, from which the given [line] $\Theta\Xi$ may be removed: therefore, the given remainder is equal in square to the rectangle $NM\Xi$, that is (by *Elements* II.6),[134] the difference of the squares on $\Xi\Theta$, ΘM: and, accordingly, the excess of the square on $\Xi\Theta$ over the square on ΞK is equal to the square on ΘM.[135] But $\Xi\Theta$, ΞK are given; and, with that, ΘM is given, and the point M is given.

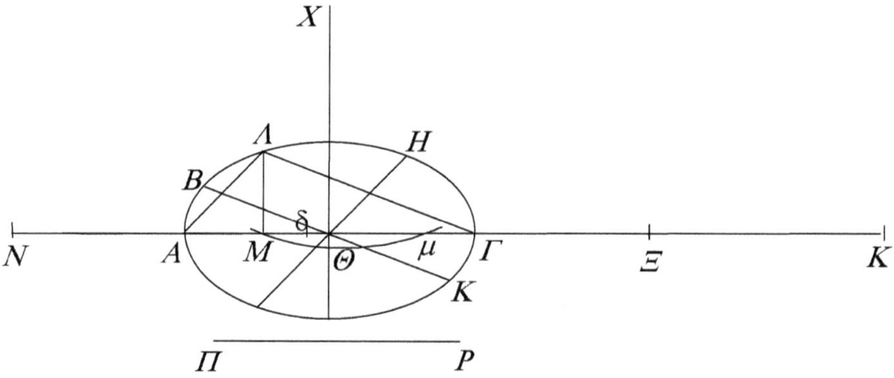

The synthesis will be accomplished in this way: let it be contrived that as the half-difference between the axis and latus rectum is to the half-sum of the conjugate [diameters] (let it be ΠP) so the half-sum will be to a fourth proportional ΘK, which is placed along the axis [extended] beyond the point Ξ. Next, let ΘX equal to $K\Xi$ be placed along the other axis, and with center X and radius $\Xi\Theta$ let the arc of a circle

[133] The argument beginning here is, *mutatis mutandis*, follows along the lines of the corresponding argument in Proposition IX.

[134] Halley must mean *Elements* II.5, which states: "If a straight line be cut into equal and unequal segments, the rectangle contained by the unequal segments of the whole together with the square on the straight line between the points of section is equal to the square on the half." In this case, the whole line is $N\Xi$, which is divided into equal parts by the center of the ellipse Θ and into unequal parts by the point M; therefore, the rectangle contained by NM, $M\Xi$ together with the square on $M\Theta$ is equal to the square on $\Theta\Xi$, that is, the rectangle contained by $NM,M\Xi$ is equal to the difference between the square on $\Theta\Xi$ and that on $M\Theta$.

[135] Rect.$NM,M\Xi$=sq.$\Theta\Xi$-sq.$M\Theta$ or sq.$M\Theta$=sq.$\Theta\Xi$-rect.$NM,M\Xi$. But rect.$NM,M\Xi$=sq.ΞK. Therefore, sq.$M\Theta$=sq.$\Theta\Xi$- sq.ΞK.

be described cutting the axis, if the problem is possible, at point M[136]; and with the line $M\Lambda$ drawn ordinatewise, let $\Gamma\Lambda$, $A\Lambda$ be joined, to which the diameters sought will be parallel. For the square on ΘM is equal to the excess of the square on $\Xi\Theta$ over the square on ΘX, or the square on $K\Xi$, exactly as was required by the analysis set out before (*ex praemissa analysi*).

But these same diameters can be found from their given sum by an entirely different method. For since (by VII.30) the sum of the squares of any conjugate diameters is equal to the sum of the square of the axis and its figure; and (by *Elements*, II.4) the square of a sum is equal to the sum of the squares of the parts together with twice the rectangle of them [*sic*], and this sum is also given: twice the rectangle contained by the conjugate diameters is now also given. But having removed twice this rectangle from the some of squares, the remainder (by *Elements*, II.7)[137] will be equal to the square of the difference of the conjugate diameters: with that, this difference is given. But since both the sum and the difference are given, also the diameters which we are seeking can themselves be obtained.

But it is necessary that the given sum of conjugate diameters be not less than the sum of the axes, nor greater than the sum of the equal conjugate diameters, or those which are equal in square to twice the squares of the axes taken together, as is demonstrated in the 26th proposition of the seventh [book].[138]

Proposition XI (Problem)

Given the axis and latus rectum of a hyperbola, find conjugate diameters whose difference equals a given line.

Let things in the hyperbola be kept as they were previously described,[139] and in the same way as was demonstrated in the preceding [propositions], it will be established that $\Gamma\Delta$ (by VII.9), or the sum of the axis and the latus rectum, is to $M\Xi$ as the square on the difference of any conjugate diameters is to the square of the difference between the same $M\Xi$ and the [line] equal in square to the rectangle

---

136 The arc will also cut the axis at another point. This point, μ, is not referred to in the demonstration but shown in the diagram.

137 *Elements* II.7 states: "If a straight line be cut at random, the square on the whole and that on one of the segments both together are equal to twice the rectangle contained by the whole and the said segment and the square on the remaining segment." The square on the remaining segment is, therefore, the square on the difference between the whole and the first segment, so that, if we take the whole to be a diameter and one of the segments to be its conjugate, then the sum of the squares on the two diameters will, by *Elements* II.7, equal the rectangle contained by the conjugate diameters together with the square on the difference of the conjugate diameters.

138 Proposition VII.26 states: "[In] every ellipse, the sum of its two axes is less than [the sum] of any conjugate pair of its diameters; and the sum of any conjugate pair of its diameters which is closer to the two axes is less than the sum of any conjugate pair of its diameters farther from the two axes; and the sum of the conjugate pair of its diameters each of which is equal to the other is greater than that of any [other] conjugate pair of its diameters." Halley's last remark follows from this propostion and VII.12. For if D and d are conjugate diameter, E is one of the equal conjugate diameters, and X and x are the axes, VII.26 tells us that $D+d \leq 2E$ or $sq.(D+d) \leq 4sq.E$. But, by VII.12, $4sq.E=2(sq.E+sq.E)=2(sq.X+sq.x)$. Therefore, for any pair of conjugate diameters, D and d, we have $sq.(D+d) \leq 2(sq.X+sq.x)$.

139 As in proposition IX.

$NM\Xi$, that is, to the square on $M\Xi$ and the rectangle $NM\Xi$ taken together, less twice the rectangle contained by $M\Xi$ and the [line] equal in square to $NM\Xi$.[140] And since $M\Xi$ is found on both sides, $\Gamma\Delta$ will be to the difference of the conjugate diameters as that same difference is to $M\Xi$, MN together (or twice ΘM), less that [line] equal in square to rectangle $NM\Xi$. If, therefore, it be contrived that $\Gamma\Delta$ be to the given difference of conjugate diameters as half that same difference is to ΘP, the line ΘP will be given, and the point P will be given. But ΘP is equal to ΘM less that [line] equal in square to rectangle $NM\Xi$; whence MP will be equal in square to the rectangle $NM\Xi$, and NM will be to MP as MP to $M\Xi$. But, *dividendo*, NP will be to PM as $P\Xi$ to ΞM, and, *permutando*, NP will be to $P\Xi$ as PM to $M\Xi$; whence, again, *dividendo*, twice ΘP will be to $P\Xi$ as $P\Xi$ to ΞM.[141] But the points Θ, P, Ξ are given. With that, the lines ΘP, $P\Xi$ are also [given]. Accordingly, the line ΞM is given, and so too the point M is given.

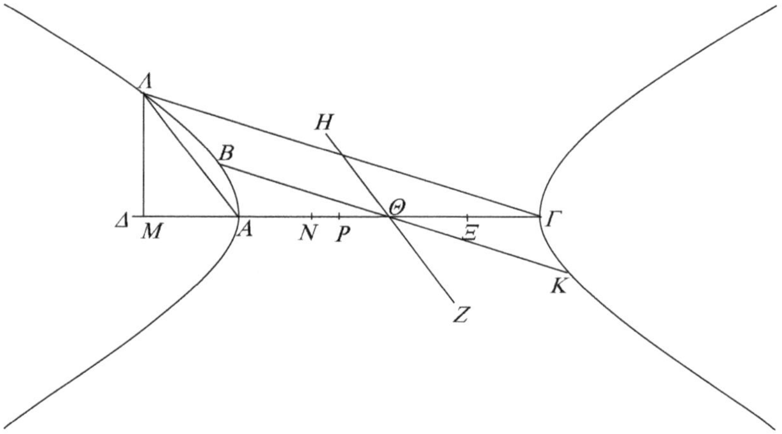

The synthesis, therefore, will be accomplished in this way. Having found N, Ξ, let it be contrived that as $\Gamma\Delta$, that is, the sum of the axis and it latus rectum, is to the given difference between the conjugate diameters, so too will the half sum of that same difference to a fourth proportional, which is ΘP situated along the axis. Next let it be contrived that as twice this same ΘP is to $P\Xi$ so too will $P\Xi$ be to ΞM, and with point M found, let the perpendicular $M\Lambda$ be set up, and let the rest be done exactly as before.

Indeed, since it was shown in the ninth [proposition] of this [work] that the rectangle contained by the sum of and the difference between any conjugate diameters of a hyperbola will be equal to the rectangle contained by the axis and the

140 Proposition VII.9 tells us that with P defined as above, namely, sq.P=rect.$MN,M\Xi$, we have sq.$A\Gamma$:sq.(BK-ZH)::rect.$N\Gamma$,$M\Xi$: sq.($M\Xi$-P). With this, or rather, sq.$A\Gamma$:rect.$N\Gamma$,$M\Xi$::sq.(BK-ZH):sq.($M\Xi$-P), the rest, as Halley points out, follows just as in proposition IX. The subsequent claims also, *mutatis mutandis*, are proven as in proposition IX.

141 "Twice $P\Theta$" because $N\Theta = \Theta\Xi$, so that $P\Xi - \mathrm{NP} = (P\Theta + \Theta\Xi) - (N\Theta - P\Theta) = (P\Theta + \Theta\Xi) - (\Theta\Xi - P\Theta) = 2P\Theta$.

difference between the axis and its latus rectum, the analogy (ἀνάλογον)[142] is that the difference between the conjugate [diameters] will be to the axis as the difference between the axis and its latus rectum is to the sum of the conjugate diameters. But everything else being given, the sum of the conjugate [diameters] is given, and the conjugate [diameters] themselves [are given], on account of the fact that both the sum and difference [of the conjugate diameters] are given.

From this, you will be able to construct both [the solution of] this eleventh problem and the ninth in a most expedient fashion.[143] For having set up normal TY upon some line ΠP, let TY be made equal to the mean proportion between the axis and the difference between the axis and the latus rectum, or that [line] equal in square to the difference between the squares of the axes, and let TP be cut off equal to difference between any conjugate [diameters], or, alternatively, ΠT [cut off] equal to their sum. Having joined PY, or $Y\Pi$, let it be divided in two at Φ, [or] X; let $\Phi\Sigma$ be drawn normal to YP, or ΣX to ΠY, which cuts line ΠP at point Σ: I say ΣP, ΣT are equal to the conjugate diameters sought, if TP was the given difference; or $\Pi\Sigma$, ΣT [are the conjugate diameters], if ΠT was the given sum of the conjugate diameters.

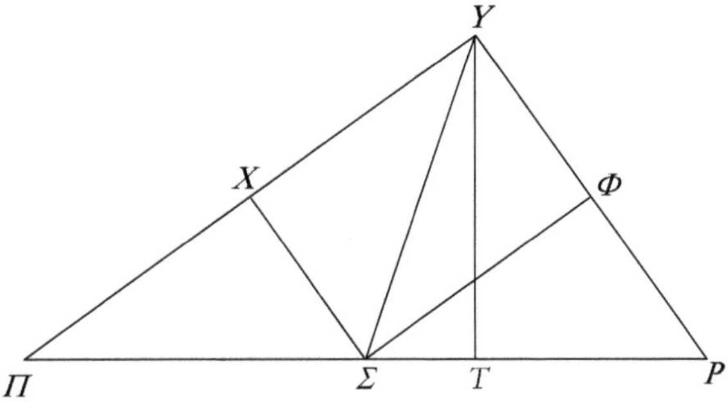

For on account of the equal lines $P\Phi$, ΦY, [the lines] ΣP, PY will be equal to one another,[144] and the square on ΣY, that is ΣP, will exceed the square on ΣT by the square on TY or the difference between the square of the axis and the figure of the section. With that (by VII.13 and 29) $P\Sigma$, ΣT are equal to the diameters sought.[145] In the same way, if ΠT is the given sum, $\Pi\Sigma$, ΣY will be equal on account of ΠY being bisected at X and the angle $\Pi X\Sigma$ being right; but the square on YT will be

---

[142] In Greek mathematics, "analogy," *analogia*, is the term for "proportion." Halley uses the Greek word (which indeed appears in Greek in Halley's text) in this way, not only here, but also in problems XX, XXIV, XXVI, XXXII, and XXXIII.

[143] *Hinc modo satis expedito tam nonum quam hoc undecimum problema contruxeris.* It is curious that Halley does not provide this procedure in the context of proposition IX—the sentence almost has the feel of an afterthought.

[144] Since $\Sigma\Phi$ is a perpendicular bisector of PY.

[145] By the Pythagorean theorem, sq.ΣY-sq.ΣT=sq.ΣP-sq.ΣT=sq.TY=sq.Axis-rect.Axis, lr.Axis. But sq.Axis-rect.Axis, lr.Axis=rect.(D-d),(D+d), where D and d are conjugate diameters. Furthermore, sq.ΣP-sq.ΣT=rect.(ΣP-ΣT),(ΣP+ΣT)=rect.TR,(ΣP+ΣT)=rect.(D-d),(ΣP+ΣT). Therefore, ΣP+ΣT=D+d. And, since also ΣP-ΣT=D-d, it follows that ΣP=D and ΣT=d.

equal to the difference of the squares on $\Pi\Sigma$, ΣT, which accordingly will be the diameters sought. It will be necessary, however, by those things demonstrated in the 27th proposition of the seventh book,[146] that the given difference between the conjugate [diameters] will be less than the difference between the conjugate axes of the hyperbola.

Corollary: From this it is manifest that, if in the most diverse hyperbolas the differences between the squares of axes and figures are equal, whatever the given magnitude of diameters you choose, equal diameters will also have equal conjugates.[147]

Proposition XII (Problem)

Given the major axis and latus rectum of an ellipse, find conjugate diameters of it whose difference is equal to a given line.

With the things we described in the ellipse kept the same, [we can say that] as the square of the axis is to the rectangle contained by $N\Gamma$,$M\Xi$ the square of the difference between the conjugate diameters will be (by VII.9) to square of the difference between the same $M\Xi$ and the [line] equal in square to the rectangle $NM\Xi$: whence by completely the same argument used in the previous [proposition],[148] as $\Gamma\delta$ (or the difference between the axis and the latus rectum) is to the difference between the conjugate diameters so too will that same difference be to $M\Xi$, MN together, that is, to $N\Xi$, or twice $\Theta\Xi$, less that [line] equal in square to rectangle $NM\Xi$. Therefore, let it be contrived that as the difference between the axis and latus rectum is to the difference between the conjugate [diameters] so too will be half of that difference to $O\Theta$, which, accordingly, will be given, as will be the point O. But $O\Theta$, as has been already said,[149] is equal to the excess by which $\Theta\Xi$ exceeds that line equal in square to the rectangle $NM\Xi$: therefore, the line $O\Xi$ is equal in square to the rectangle $NM\Xi$,[150] that is (by *Elements* II.6),[151] the difference between the squares on $\Theta\Xi$, ΘM; and, with that, the square on ΘM is equal to the excess by which the square on $\Theta\Xi$ is greater than the square on $O\Xi$. But $\Theta\Xi$, $O\Xi$ are given, whence also ΘM is given, and the point M is given.

The synthesis of the problem will be accomplished, therefore, if it be contrived that as the difference between the axis and the latus rectum to the given difference

---

146 "[In] every ellipse, or hyperbola in which the two axes are unequal, the increment of the greater axis over the lesser is greater than the increment of [the greater of] any conjugate pair [sic] of its diameters over the diameter conjugate with it; and the increment of [the greater of a pair of] them nearer to the greater axis over the diameter conjugate with it is greater than the increment of [the greater of a pair of them] farther [from the major axis] over the diameter conjugate with it."

147 This already follows from VII.13 (or VII.29). Halley probably wants to bring out that the construction given at the end of VIII.11 does not refer to any particular hyperbola and, therefore, to all hyperbolas.

148 Previous proposition, first paragraph.

149 See the previous proposition with $P\Theta$ playing the part of $O\Theta$ and ΘM playing the part of $\Theta\Xi$.

150 Again with P defined so that sq.P=rect.MN,$M\Xi$, we have $O\Theta=\Theta\Xi-P$ or $O\Xi=\Theta\Xi-O\Theta=P$; so, sq.$O\Xi$=rect.NM,$M\Xi$.

151 See note to proposition X above.

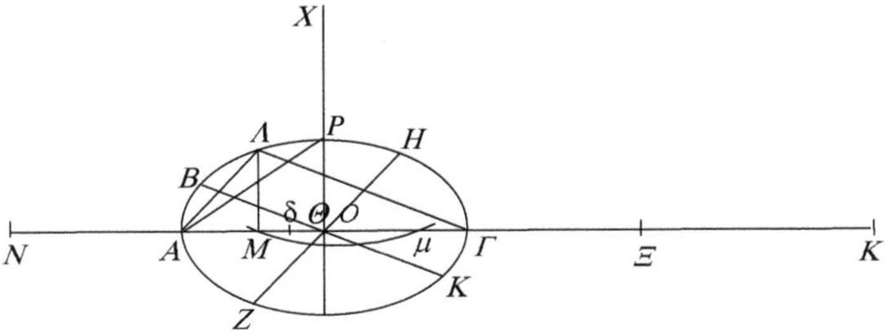

between the conjugate diameter so too will half the given difference be to ΘO, which lies along the axis from the center O towards Ξ. Next, let ΘX, in the extension of the minor axis, be made equal to this same ΘO, and with center X and radius $\Theta\Xi$ let the arc of a circle be described cutting the axis in points M, μ, at an equal distance from the center Θ; and with a perpendicular such as $M\Lambda$ set up, the diameters sought will be found as described above. The reasoning behind this synthesis is clear from the analysis and from *Elements* I.47.[152]

But the proposed difference must not be greater than the difference between the axes of the ellipse, for (by VII.27) if it is set greater than that, the problem will be impossible, and the point M will lie outside the axis.

But we can find the conjugate diameters of the ellipse given their sum or difference, in another, and not at all inelegant, way. For since AP (or that equal in square to squares of the semi-axes taken together) is equal in square also (by the twelfth proposition of Book VII) to the squares of any semi-conjugate diameters whatsoever,[153] let circle $A\Gamma B\Delta$ be described having radius AP, and from the center P let a perpendicular $\Gamma P\Delta$ be set up above the diameter AB cutting the circle at Γ, Δ. Next, with center Δ and radius ΔA let arc AZB be curved about (*circinetur*), which on account of the equal [lines] AP, $P\Delta$ will be a quarter circle, and, accordingly, the angle that it encloses (*capit*) will be (by *Elements* III.20)[154] one and a half times a right angle (*sesquialter anguli recti*). In this arc, let ZB, equal to the difference between the conjugate diameters, be inscribed, and let it be produced until it cuts the semi-circle $A\Gamma B$ at point H: I say lines BH, HZ are equal to the conjugate diameters being sought.

Indeed, let AH, AZ be joined; AHB will also then be right (by *Elements* III.31)[155]; but the HZA, which is adjacent to AZB, is half of a right angle, and with that, angle

[152] The "Pythagorean Theorem."

[153] That is, to the sum of the squares of any pair of conjugate semi-diameters, or four times the sum of the squares of the conjugate diameters themselves.

[154] *Elements* III.20 states: "In a circle the angle at the center is double of the angle at the circumference, when the angles have the same circumference as base." In this case, since angle $A\Delta B$ subtending the arc $A\Delta B$ is three right angles, the angle at the circumference, AZB is half of three right angles.

[155] *Elements* III.31 states: "In a circle the angle in the semicircle is right, that in a greater segment less than a right angle, and that in a less segment greater than a right angle; and further the angle of the greater segment [i.e., the angle between one side of the angle and the arc of the circle] is greater than a right angle, and the angle of the less segment less than a right angle."

HAZ is also equal to angle *HZA*, whereby also *AH* is equal to *HZ*. Therefore, the squares on *AH* and *HB* (that is, *ZH, HB*) taken together, on account of the right angle *AHB*, are equal to four times the square on *AP*, that is, by construction, to the squares of the axes of the ellipse taken together. The difference of the same lines, *BH, HZ*, is the proposed line *ZB*,[156] whence (by VII.12) *BH, HZ* are equal to the conjugate diameters sought.[157]

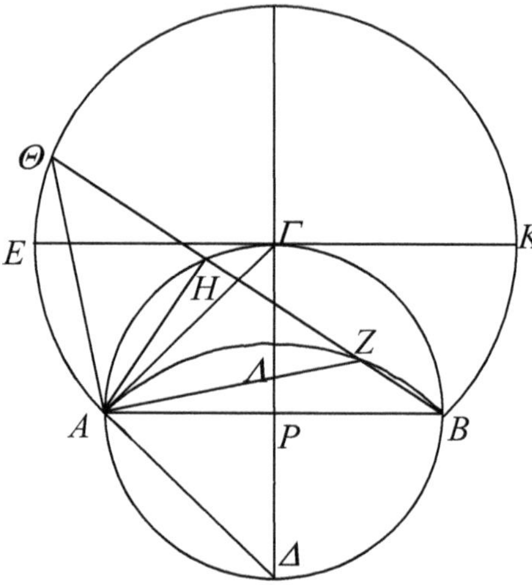

If, however, as in proposition 10, the sum of the conjugate [diameters] be given, and it be proposed to exhibit those diameters, then let arc *AΘB* be described with center *Γ* and radius *ΓA*; this arc will enclose (by *Elements* III.20) half a right angle.[158] Therefore, if the line equal to the sum of the conjugate [diameters], suppose it be *BΘ*, be inscribed in the arc [*AΘB*] and *AΘ, AH* be drawn, then the angle *AΘH* will be half a right angle. And, on account of the right angle *AHB*, angle *ΘAH* will also be half a right angle, and, accordingly, *ΘH* will be equal to *HA*. But the squares on *AH, HB*, that is, on *ΘH, HB*, are equal to the square on *AB* or the squares of the axes [of the ellipse] taken together; whence the lines *ΘH, HB*, whose sum is *ΘB*, are equal to the conjugate diameters which we were required to find.

Corollary: But it is clear that if the sums of the squares of the axes in any ellipses of different types all be equal to one another, then whatever diameters given in

[156] The line set out equal to the given difference.

[157] For if the sum of the squares on the axes of the ellipse is K^2, then the locus of points *H*, such that $AH^2 + HB^2 = K^2$ is the semicircle *AHB* (in fact, the entire circle—but it suffices, by symmetry, to consider only the semicircle). To each point *H*, however, there corresponds a unique point *Z* on the arc *AZB*, and, therefore, a unique length *BZ*. Hence, given *BZ*, *H* is uniquely determined, and, therefore, also *AH* and *HB* are uniquely determined.

[158] Since angle *AΓB*, at the center, is right.

magnitude you choose, those diameters with equal conjugate [diameters] will also be equal.[159]

It ought to be noted, in passing, that as the square on the radius AP is equal to *Hippocrates' lune* $A\Lambda B\Gamma$, if diameter $E\Gamma K$ is drawn parallel to AB, the area $AEKB\Lambda$ will equal the square on $E\Gamma$, and, accordingly, to twice the lune $A\Lambda B\Gamma$: whence area $E\Gamma HA$ will be equal to half the lune $AH\Gamma\Lambda$. The demonstration of this fact is clear.[160]

Proposition XIII (Problem)

Given the axis and latus rectum of a hyperbola, find both the magnitude and position of conjugate diameters which contain a rectangle equal to a given rectangle.

With the figure of the hyperbola kept as described before, suppose BK, ZH are the diameters sought. Since, however, (by VII.10) the square on $A\Gamma$ is to the rectangle contained by the conjugate diameters of the hyperbola as $N\Gamma$ is to that equal in square to the rectangle $NM\Xi$,[161] and (by the demonstration of VIII.7) the rectangle contained by $N\Gamma$ and the sum of the axis and latus rectum is equal to the square on $A\Gamma$,[162] eliminating $N\Gamma$ from each (*sublato utrinque*), the rectangle contained by $\Gamma\Delta$ (the axis and its latus rectum taken together) and that equal in square to the rectangle $NM\Xi$ will be equal to the given rectangle contained by the diameters sought, BZ, ZH.[163]

---

[159] If the sum of the squares of the axes is given, say, equal to AB^2, the conjugate diameters must be pairs of lines such as AH, HB, where H is a point on the semicircle AHB having diameter AB. So, *regardless of the latus rectums of the ellipses*, if the sums of squares of the axes in two ellipses are both AB^2 and a diameter in one is equal to a diameter in the other—equal, say, to BH—then the conjugates of each must be equal to AH.

[160] See appendix 1.

[161] This is the same as Apollonius's enunciation of VII.10, which is also given in terms of the previous figure, that is, there is no additional *protasis*.

[162] Why Halley cites VIII.7 here is unclear. Indeed, the fact that rect.$N\Gamma$,(axis+latus rectum)=sq.$A\Gamma$ (=sq.axis) follows from the definition of the homologue: $AN : N\Gamma$::latus rectum: axis; hence, $(AN + N\Gamma)$:$N\Gamma$::(axis+latus rectum): axis or $A\Gamma$:$N\Gamma$::(axis+latus rectum):$A\Gamma$, from which the result is immediate.

[163] The words "*sublato utrinque* $N\Gamma$" here, as in Prop. IX, betray Halley's use of algebra to work out some of his results. In this case, the algebra must have been something like this: sq.$A\Gamma$/rect.$(BZ,HZ) = N\Gamma/P$ (sq.P =rect.$NM\Xi$) and rect.$N\Gamma,\Gamma\Delta$=sq.$A\Gamma$; therefore, rect.$(N\Gamma,\Gamma\Delta)$/rect.$(BZ,HZ) = N\Gamma/P$, so that, "eliminating $N\Gamma$ from both," we have $\Gamma\Delta$/rect.$(BZ,HZ) = 1/P$ (an expression impossible to write in an Apollonian context!), from which we obtain rect.BZ,HZ =rect.$\Gamma\Delta,P$. Apollonius could have drawn the same conclusion as follows: since (reverting to the usual signs for ratios and proportions) sq.$A\Gamma$:rect.(BZ,HZ)::$N\Gamma : P$, also, with $\Gamma\Delta$ as a common side, sq.$A\Gamma$:rect.(BZ,HZ)::rect.$N\Gamma,\Gamma\Delta$:rect.$P,\Gamma\Delta$; but rect.$N\Gamma,\Gamma\Delta$=sq.$A\Gamma$; therefore, sq.$A\Gamma$:rect.(BZ,HZ)::sq.$A\Gamma$:rect.$P,\Gamma\Delta$ from which we have the result. Halley could have framed this kind of argument easily enough. That he did not, suggests that such quasi-algebraic formulations were not, to his mind, violations of his otherwise historically faithful approach.

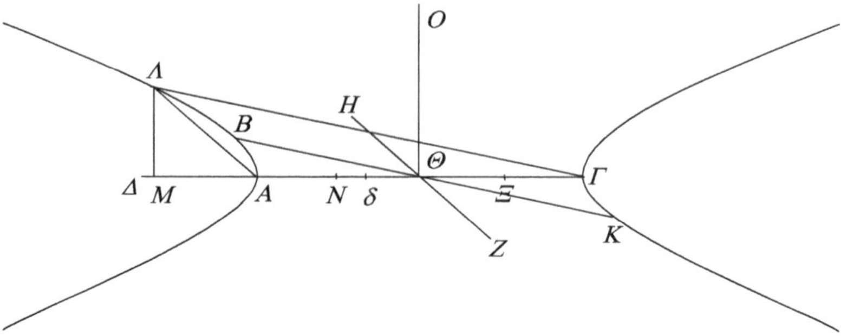

If, therefore, the given rectangle is applied to $\Gamma\Delta$,[164] the sum of the axis and its latus rectum, its width will be given, namely, that line equal in square to the rectangle $NM\Xi$, that is, (by *Elem.* II.6) that line equal in square to the difference of squares on ΘM, $M\Xi$. Let that width be ΘO, and the square on ΘO will be equal to the difference of squares on ΘM, $\Theta\Xi$: add the square on $\Theta\Xi$ to each side and the squares on ΘO and $\Theta\Xi$ will be equal to the square on ΘM. But ΘO, $\Theta\Xi$ are given, and with that, ΘM is given, whence also point M is given.

And so, the synthesis will be accomplished in this way. Keeping everything in the previous figure of the hyperbola the same, apply the given rectangle contained by the conjugate diameters to the line $\Gamma\Delta$, or to that equal to the sum of the axis and its latus rectum; let the width be the line ΘO, so that the rectangle contained by $\Gamma\Delta$, ΘO be equal to the given rectangle; let ΘO be placed along the conjugate axis and let $O\Xi$ be joined, and let ΘM be made equal to $O\Xi$; M now having been found, let the rest be done as before. The demonstration is clear from the analysis and from *Elem.* I.47.

Proposition XIV (Problem)

*Similarly for an ellipse, given the axis and latus rectum, it is required to find both the magnitude and position of conjugate diameters of [the ellipse], which take in (*comprehendatur*) a rectangle equal to a given rectangle.*

With the figure of the ellipse kept as described before, and with the same argument we used before, it will be the case that (by the tenth proposition in Book VII) as the rectangle contained by $N\Gamma$ and $\Gamma\delta$, the difference between the axis and its latus rectum (that is, the square on $A\Gamma$),[165] is to the rectangle contained by the conjugate diameters,

---

[164] This is a technical term in Greek mathematics: *to apply* (παραβάλλειν) an area A to a line m means to find a line n so that area A=rect.m,n. In the *Elements*, the idea of application of areas first appears in I.44; in the *Conics*, the three basic kinds of application of areas—*parabolic*, *elliptic*, and *hyperbolic*—are connected to the characteristic properties, the *symptōmata*, of the conic sections and account for the names Apollonius gives to the sections.

[165] By the definition of the *homologue* for the ellipse, $N\Gamma$:NA::axis: latus rectum::$A\Gamma$:$A\delta$; therefore, *convertendo*, $N\Gamma$:$N\Gamma$-NA::axis: axis – latus rectum, or $N\Gamma$:$A\Gamma$::$A\Gamma$:$\Gamma\delta$, so that rect.$N\Gamma$, $\Gamma\delta$=sq.$A\Gamma$.

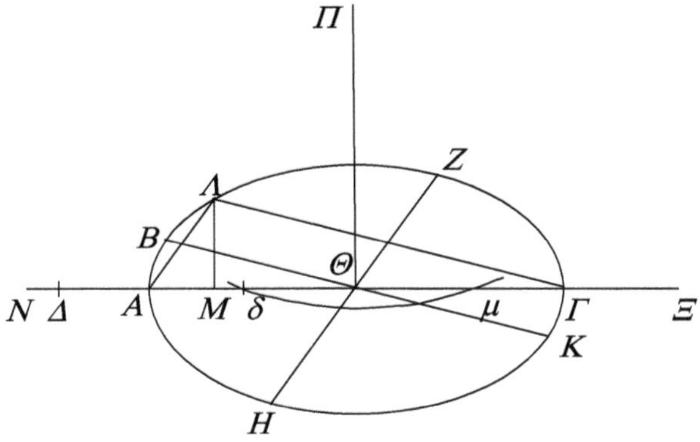

BK, ZH, so too will $N\Gamma$ be to that [line] equal in square to the rectangle $NM\Xi$. And on account of $N\Gamma$ being found on both sides, the rectangle contained by $\Gamma\delta$, the difference between the axis and its latus rectum, and that [line] equal in square to the rectangle $NM\Xi$ will be equal to the given rectangle,[166] namely, that contained by the conjugate diameters sought, BK, ZH. Therefore, applying that rectangle[167] to the given line $\Gamma\delta$, the width, which is derived from the application (*ex applicatione orta*), will be given [and] equal to that line equal in square to $NM\Xi$, that is (by *Elements* II.5), to the line equal in square to the excess by which the square on $\Theta\Xi$ exceeds the square on ΘM.[168] Let this width be equal to the line $\Theta\Pi$, so the square on $\Theta\Pi$ will be excess by which the square on $\Theta\Xi$ exceeds the square on ΘM. Therefore, the square on $\Theta\Xi$ will exceed the square on $\Theta\Pi$ by the square on ΘM. But $\Theta\Xi$, $\Theta\Pi$ are given: whence the line ΘM is also given and the point M is given.

And so, the synthesis is clear. For if the given rectangle is applied to $\Gamma\delta$, the difference between the axis and its latus rectum, that is, if the width $\Theta\Pi$ is obtained (*habeatur*) so that the rectangle contained by $\Theta\Pi$, $\Gamma\delta$ be equal to the given rectangle, and $\Theta\Pi$ is set along the minor axis extended; and afterwards [if] a circular arc with radius $\Theta\Xi$ and center P is described cutting the axis at points M, μ and a perpendicular such as $M\Lambda$ is set up, [then] we shall have, in the manner we have stated many times before, the conjugate diameters whose rectangle[169] will be equal to the given one.

---

[166] Again, the argument is algebraic: referring, thus, to products rather than rectangles, we have $(N\Gamma\times\Gamma\delta)/(BK\times ZH) = N\Gamma/P$ (sq.P=rect.$NM\Xi$) or, "since $N\Gamma$ is found on both sides," $\Gamma\delta/(BK\times ZH)=1/P$, from which we have $\Gamma\delta\times P = BK\times ZH$.

[167] Namely, the rectangle contained by the conjugate diameters.

[168] See note on proposition X.

[169] That is, the rectangle contained by the conjugate axes.

It will be necessary, however, that (by VII.28)[170] the given rectangle is not less than the rectangle contained by the axes; nor can it be greater than the square contained by equal diameters,[171] that is (by VII.12) half the sum of the squares on the axes, or the rectangle contained by $\Theta\Gamma$, $\Gamma\Delta$, which is equal to half the sum of the square of the axis and its figure.[172]

We can find, however, these same diameters in an entirely different way. For since (by VII.12) the sum of the squares on *BK, ZH* be equal to the sum of the squares on the axes of the ellipse, if to that sum is added twice the rectangle on *BK,ZH*, it will produce (by *Elements* II.4) the square on *BK, ZH* taken together, and, with that, *BK, ZH* together will be given.[173] Indeed, if from the same sum of squares twice that rectangle is removed, there will remain (by *Elements* II.7) the square of the difference of those [same diameters] *BK, ZH*, and, with that, this difference is given. But the sum and difference of the two lines being given, the lines themselves will also be given.

Hence, synthesis will be accomplished if *AB* is [equal] in square to the sum of the squares of the axes; and $B\Gamma$, which is [equal] in square to twice the given rectangle [contained] by the sought diameters, is placed at a right angle [to *AB*]; and an arc of a circle ΓE with center *B* and radius $B\Gamma$ is described. Let the line *AB* be bisected at Δ, and let a small part of a semicircle with center Δ and radius $B\Delta$ be described cutting the arc ΓE at *E*, and let $AE, A\Gamma$ be joined. These, by what has already been said, will be equal to the sum and difference of the diameters sought. Let *AZ* be made equal to *AE* and let $Z\Gamma$ be bisected at *H*; then *AH* will be the greater of the diameters and $H\Gamma$ the less.[174]

---

[170] *Conics* VII.28 states: "[In] every hyperbola or ellipse, the rectangle formed by the product of its two axes is less than the rectangle formed by the product of any conjugate pair whatever of its diameters; and, of the conjugate diameters, for those in which the longer [of the pair] is closer to the greater axis, the product of [that diameter] and the diameter conjugate with it is less than the product of one of those in which it is farther from it [the greater axis] and the diameter conjugate with it." I ought to mention here that although I am not competent to judge whether "product" is an accurate translation of the Arabic of the Banū Mūsā text, I feel confident to say that it is highly unlikely that this was an accurate translation of the Greek text, which was almost certainly a variation of the phrase τὸ περιεχόμενον ὀρθογονιον ὑπὸ....

[171] This follows, of course, from the second part of VII.28.

[172] Recall the figure is the rectangle contained by the axis and its latus rectum: since $\Gamma\Delta$ equals the axis together with its latus rectum and $\Theta\Gamma$ is half the axis, the rectangle in question equals ½(sq.Axis+rect.Axis,latus rectum). Moreover, by *Conics* I.15, the figure on any diameter equals the square on the conjugate diameter (see note to Prop. VI above); therefore, ½(sq.Axis+rect.Axis,latus rectum)=½(sq.Axis+sq.conj.Axis).

[173] That is, the sum of *BK* and *HZ*.

[174] The argument is as follows. Let the diameters sought be D and d, where D is the larger of the two. First, sq.*AB* =sq.D+sq.d=sum of the squares on the axes, and sq.$B\Gamma$=2rect.D,d, which is given. Now, the point *E* lies on a circle whose diameter is *AB*; therefore, triangle *AEB* is right and sq.*AE*+sq.*BE* =sq.*AB* or sq.*AE* =sq.*AB*−sq.*BE*. But $BE = B\Gamma$; therefore, sq.*AE* =(sq.D+sq.d)−2rect.D,d=sq.(D−d); therefore, *AE* =D−d. But sq.$A\Gamma$=sq.*AB*+sq.$B\Gamma$=(sq.D+sq.d)+2rect.D,d=sq.(D+d), so that $A\Gamma$=D+d . Hence, with $AZ = AE$ =D−d and $A\Gamma$=D+d, *AH* =½($AZ+A\Gamma$) =½(2D)=D and $H\Gamma = A\Gamma - AH$ =(D+d)−D=d.

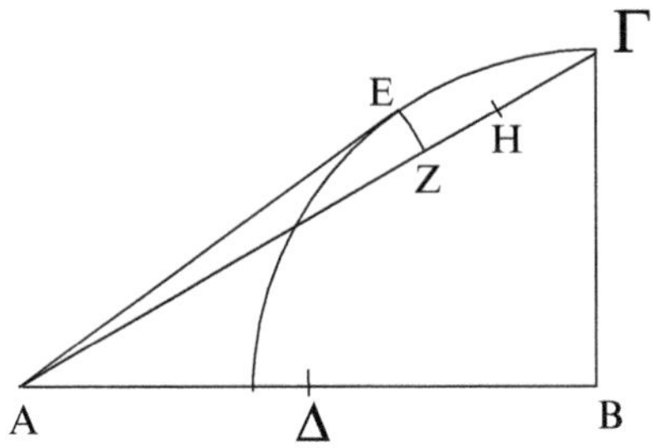

Proposition XV (Problem)

Given the axis and latus rectum of a hyperbola, find the position and magnitude of conjugate diameters the sum of whose squares is given.

With the figure of the hyperbola kept as described before, and having set out the sum (*aggregato*) of some [pair] of conjugate diameters, it is required to find those diameters. Since (by VII.11) the square on $A\Gamma$ is to the sum of the squares on the conjugate diameters of the hyperbola as $N\Gamma$ is to NM, $M\Xi$ taken together,[175] and the square on $A\Gamma$ is equal to the rectangle contained by $N\Gamma$ and $\Gamma\Delta$, the sum of the axis and its latus rectum, [then] on account of $N\Gamma$ common to both, the rectangle contained by $\Gamma\Delta$ and NM, $M\Xi$ together, or twice ΘM, is equal to the given sum of squares on the conjugate diameters.[176] Applying, therefore, the given sum of these squares to $\Gamma\Delta$, the sum of the axis and its latus rectum, the width which is equal to twice ΘM will be given. Therefore ΘM is given, and because Θ is given, the point M is also given.

And so the synthesis will be accomplished, if the half-sum of the squares on the conjugate diameters is applied to $\Gamma\Delta$, or to the sum of the axis and its latus rectum, and the width of the application[177] is placed along the axis from the center Θ to the point M: the point M having been found, the diameters will be obtained exactly as before. The demonstration is manifest from the analysis.

It is furthermore manifest that the sum of the proposed squares can not be less than the sum of the squares of the axes,[178] that is, the sum of the squares on the [principal] axis and its figure or the rectangle $A\Gamma\Delta$.

---

[175] *Conics* VII.11 states, just as Halley presents it, that sq.$A\Gamma$:(sq.BK+sq.ZH)::ΓN:(NM+$M\Xi$). Like proposition VII.10, proposition VII.11 contains no *protasis* and refers back to the diagrams in VII.6,7, which are similar to Halley's diagrams here and in the previous proposition.

[176] As we have already seen, the argument, introduced by the phrase "$N\Gamma$ being on both sides," is plainly algebraic. Accordingly, writing the claims as products and quotients, we have $A\Gamma^2/(BK^2+ZH^2)=N\Gamma/(NM+M\Xi)$, or since $A\Gamma^2=N\Gamma\times\Gamma\Delta, N\Gamma\times\Gamma\Delta/(BK^2+ZH^2)=N\Gamma/(NM+M\Xi)$, or, since $N\Gamma$ is on both sides, $\Gamma\Delta/(BK^2+ZH^2)=1/(NM+M\Xi)$, from which the result follows.

[177] That is, the width of the rectangle applied to $\Gamma\Delta$ which is equal to half the sum of the squares.

[178] This is stated explicitly in the first corollary following *Conics* VII.31.

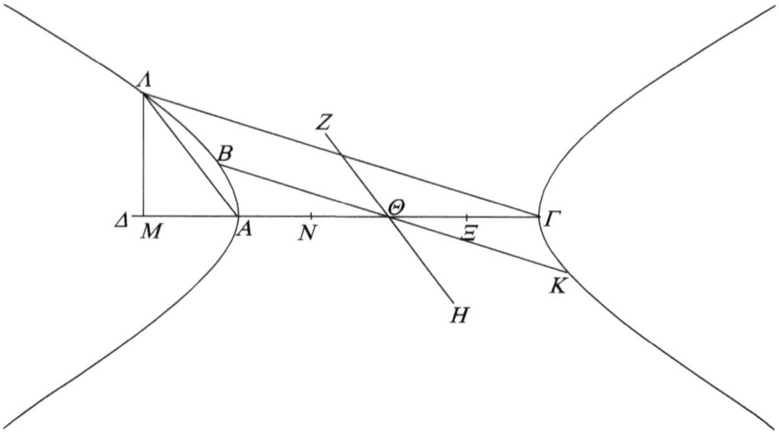

Proposition XVI (Problem)

Given the axis and latus rectum of an ellipse, it is required to find the position and magnitude of conjugate diameters whose squares have a given difference.

With the figure of the ellipse being as described before, let *BK, ZH* be the conjugate [diameters] whose squares have a given difference; we shall investigate the position and magnitude of these in this way. Since (by VII.14) the square on $A\Gamma$ is to the difference of the squares on the conjugate diameters as $N\Gamma$ is to twice ΘM[179], [then] by the argument we have taken advantage of so many times,[180] the rectangle contained by $\Gamma\delta$ (the difference between the axis and its latus rectum) and twice ΘM will equal the difference of the squares on the conjugate diameters: and so, applying the given difference of squares to $\Gamma\delta$ (the given difference between the axis and latus rectum), the given width emerges, namely, [the line equal to] twice ΘM: therefore, ΘM is given and so the point M is given as well.

The synthesis, moreover, is clear: for if the half-difference of the squares on the conjugate diameters is applied to $\Gamma\delta$, the difference between the axis and its latus rectum, the width equal to the sought line ΘM will appear, whence the other things will follow suit.

It is necessary though that the difference of the proposed squares be not greater than the difference of the squares of the axes,[181] or the difference of the square and figure of the axis, that is, the rectangle, $A\Gamma\delta$.

We can solve the XVIth problem and, equally, the XVth problem differently with the help of the 12th and 13th propositions from Book Seven. For since in the case of the hyperbola (by VII.13) the difference of the squares of the axes is equal to the

---

179 This again is very nearly Apollonius's own enunciation of *Conics* VII.14.

180 Namely, from the basic proportion $N\Gamma:NA::A\Gamma:A\delta$ we derive equality sq.$A\Gamma$=rect.$N\Gamma,\Gamma\delta$; then, by VII.14 and with Δ indicating the difference of the squares on the conjugate diameters, we write $N\Gamma:2\Theta M$::sq.$A\Gamma:\Delta$::rect.$N\Gamma,\Gamma\delta$: Δ; then "$N\Gamma$ being on both sides," we obtain (now writing the expression algebraically as before) $1/2\Theta M=\Gamma\delta/\Delta$, so that Δ=rect. $\Gamma\delta,2\Theta M$.

181 By the third corollary after *Conics* VII.31.

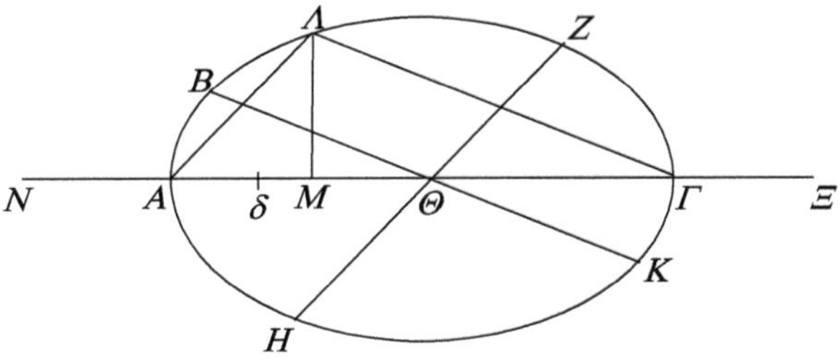

difference of the squares of any conjugate diameters, if the half-sum of the proposed sum of squares on the same [conjugate diameters] is added and subtracted from the given half-difference, then each of the two diameters sought will be given being equal [respectively], naturally, to the sum and the difference of the givens. Equally, on account of the given sum of the squares (by VII.12), if the difference is also given, the square of each diameter can be found by the same argument.[182] Whence, if it were desired, we could also find the point M as in the demonstration of the 5th and 6th [problems] in Book VIII.

Proposition XVII (Problem)

Given the axis and latus rectum of a hyperbola, find the magnitude and position of conjugate diameters containing a given angle.

The solution to this problem, as well as to the next, is provided by Apollonius at the end of Book II.[183] There, however, he assumes the sections are already described. Among *diorism* theorems[184] in the seventh book, proposition 31 seems to

---

[182] If D and d are conjugate diameters and X and x are the axes of a hyperbola, then *Conics* VII.13 tells us that sq.D-sq.d=sq.X-sq.x. Suppose the sum of squares sq.D+sq.d is given. Then 1/2(sq.D+sq.d)±1/2(sq.X-sq.x) are given. But 1/2(sq.D+sq.d)+1/2(sq.X-sq.x)=1/2 (sq.D+sq.d)+1/2(sq.D-sq.d)=sq.D. Therefore, sq.D and, accordingly, D is given in magnitude. Similarly, 1/2(sq.D+sq.d)-1/2(sq.X-sq.x) =1/2(sq.D+sq.d)-1/2(sq.D-sq.d)=sq.d. So, d is given in magnitude. *Conics* VII.12 tells us that, for the ellipse, sq.D+sq.d=sq.X+sq.x. Then since, in VIII.16, sq.D-sq.d is given, 1/2(sq.D-sq.d)±1/2(sq.X+sq.x)=1/2(sq.D-sq.d)+1/2(sq.D+sq.d) are given, and, again, we obtain D and d for the ellipse.

Implicitly, this whole development is related to Pappus's lemma 8 for *Conics* VII-VIII: "Let the sum of the squares on AB and *BΓ* be given and also the excess of the squares on AB and *BΓ*: [I say] that each of AB and *BΓ* is given" (*Collection*, VII, Hultsch, p. 996). This is one of the few places in Halley's reconstruction where one can point to a connection between Halley's development and Pappus's lemmas. Given the remarks in Halley's introduction about the importance of these lemmas for his reconstruction, it is curious that Halley does not mention Pappus explicitly here.

[183] Halley must have in mind *Conics* II.51: "Given a section of a cone, to draw a tangent which with the diameter drawn through the point of contact will contain an angle equal to a given acute angle." Since the tangent is parallel to the conjugate diameter, *Conics* II.51 allows us, alternatively, to find a pair of conjugate diameters containing a given angle.

[184] Recall, this is how Halley refers to Book VII in the letter to Henry Aldrich opening Book VIII and how Apollonius, himself, refers to Book VII in the letter introducing Book I of the *Conics*.

I might also remark that although the word *diorism* generally appears in Greek in the text, in this spot Halley writes in Greek-inflected Latin, *Theoremata dioristica*.

have been inserted to pave the way for this very problem when it has been assumed that the curve has not yet been described, as stated in the premises.[185]

For since (by VII.31) the rectangle contained by the axes is equal to the non-rectangular (*obliquangulo*) parallelogram enclosed by any two conjugate diameters, if a perpendicular is dropped from the extremity of one of the diameters to its conjugate, twice the rectangle contained by the perpendicular and the conjugate diameter will equal the rectangle contained by the axes of the section.[186] Accordingly, that rectangle [contained by the perpendicular and conjugate diameter] will be to the rectangle contained by the conjugate diameters as the perpendicular is to the semi-diameter from whose extremity the perpendicular was dropped.[187] But the angle being given, this ratio is given, and, with that, because the rectangle contained by the axes [is given], the rectangle contained by the conjugates is given.

Assume now that what we are trying to do has been done, and let BK, ZH be the conjugate diameters containing the angle $B\Theta Z$ equal to the given angle. Let the perpendicular BT be dropped [from the vertex B] to diameter ZH. On account of the given angle $B\Theta Z$, the ratio BT to $B\Theta$ will be given. But, by what has been already said [above], as BT is to $B\Theta$ so too is the rectangle contained by the axes to the rectangle contained by BK, ZH. And, on account of the given axes, the rectangle contained by BK, ZH will also be given. But, when the rectangle contained by the conjugate diameters are given, the diameters themselves are also given both in magnitude and position by what we demonstrated in the 13th [proposition] of this [book].

From here the synthesis can be accomplished thus. Let the angle $A\Theta\Pi$ be made equal to the given angle. Let ΘP be cut off [so that] it is to the axis $A\Gamma$ as the conjugate axis is to $\Gamma\Delta$ or the sum of the axis and its latus rectum. Let PX be erected at a right angle from [*super*] ΘP and let it intersect the extended conjugate axis at X.[188] I say $X\Xi$, supposing it to have been joined, is equal to ΘM.[189] With point M having been found, however, the perpendicular ΛM can be erected, and we shall have the rest as before.

Indeed, we contrived the rectangle contained by $\Gamma\Delta$, ΘP to be equal to the rectangle contained by the axes; and because the angle ΘXP is equal to the angle $Z\Theta B$,

---

185 *Conics* VII.31 states: "When a pair of conjugate diameters is drawn in an ellipse or between conjugate opposite sections [i.e., hyperbolas], then the parallelogram bounded by that pair of diameters with angles equal to the angles formed by the diameter at the center is equal to the rectangle bounded by the two axes."

186 By *Elements* I.35, and because the perpendicular, having been dropped from the *semi*-diameter, is half the altitude of the parallelogram contained by the conjugate diameters.

187 Let D, d be the conjugate diameters, N the perpendicular dropped from the extremity of D to d. Then rect.N,d:rect.½D,d::N:½D by *Elements* VI.1. The latter ratio is given because the angle contained by D and d is given.

188 Hence, it is angle $PX\Theta$ which is equal to the given angle.

189 Once it is demonstrated that the rectangle contained by $\Theta X,\Gamma\Delta$ is equal to that contained by BK,ZH (as Halley goes on to do in the next paragraph), the reasoning can proceed exactly as in VIII.13 above with ΘX, here, playing the part of ΘO: thus, again, sq.ΘM will be shown equal to sq.$X\Theta$+sq.$\Theta\Xi$ or sq.$X\Xi$.

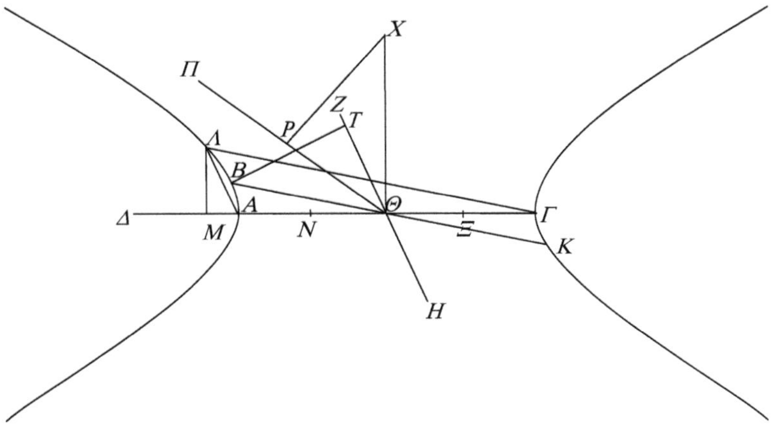

$P\Theta$ will be to ΘX as TB is to $B\Theta$, that is (by VII.31), as the rectangle contained by the axes is to the rectangle contained by the diameters BK, ZH. Therefore the rectangle contained by $\Gamma\Delta, \Theta X$ is equal to that contained by the conjugate diameters, BK, ZH; accordingly, by what was laid out in the synthesis of problem 13, the point M has been properly (*rite*) found. And it is clear that this angle is not bound (*non habere limitem*); but where the conjugate diameters are closer to the asymptotes [the angle] will, to that extent, come out smaller.

Proposition XVIII (Problem)

Similarly, in an ellipse, given the axis and its latus rectum, it is required to find the magnitude and position of conjugate diameters containing a given angle.

The rectangle contained by the axes of the ellipse is equal to any non-rectangular parallelogram contained by conjugate diameters (by the same VII.31): accordingly, having dropped a perpendicular, HT, from the extremity of one of the diameters ZH to its conjugate BK, twice the rectangle contained by BK, HT will be equal to the rectangle contained by the axes. [But] the latter is given, and, with that, the rectangle contained by BK, HT is given. But the rectangle contained by BK, HT is to the rectangle contained by BK, $H\Theta$ as HT is to $H\Theta$. [190] But because the angle $B\Theta H$ is given the ratio HT to $H\Theta$ is given. Accordingly, the rectangle contained by BK, $H\Theta$, twice of which is the rectangle contained by BK, HZ or the rectangle contained by the diameters sought, is given. The rectangle contained the diameters being given, however, the line ΘM will also be given (by VIII.14); whence the point M [will be] given.

The synthesis of the problem will thus be accomplished in the following way. Let angle $A\Theta P$ be made equal to the given angle contained by the conjugate diameters, and let ΘP be cut off so that the rectangle contained ΘP and $\Gamma\delta$ (the difference between the axis and its latus rectum) be equal to the rectangle contained by the axes of the section; and let the perpendicular $P\Pi$ be set up intersecting the extended

[190] The argument is the same as that in the second paragraph of the previous proposition.

minor axis at Π; next, with center Π and radius $\Theta\Xi$, let a circular arc be described cutting the major axis at points M, μ; and let perpendiculars [*sic*.] such as $M\Lambda$ be set up, from which the rest follows in the way discussed so many times before.

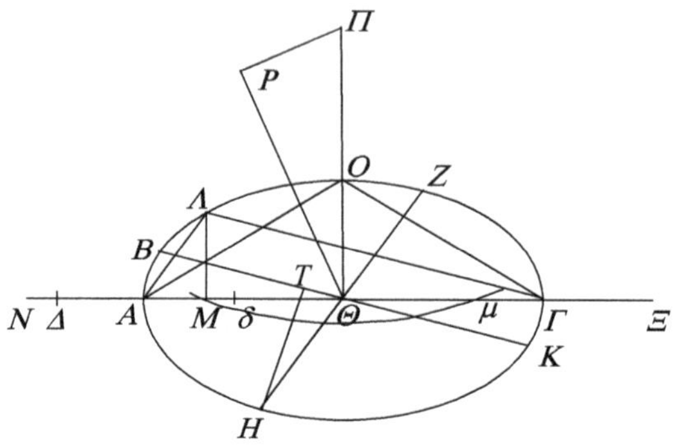

For the rectangle contained by $\Gamma\delta$, ΘP is equal to the rectangle circumscribed about the ellipse, which is to the rectangle contained by the conjugate [diameters] BK, ZH as HT is to $H\Theta$, that is, as ΘP is to $\Theta\Pi$, because the angle $\Theta\Pi P$ has been made equal to the angle $B\Theta H$.[191] Accordingly, the rectangle $\Gamma\delta$, $\Theta\Pi$ will be equal to the rectangle contained by BK, ZH : whence (by what was found in the 14th [proposition] of this [book]), the circle described with center Π and radius $\Theta\Xi$ will necessarily pass through the point M.

It is necessary, however, that the acute angle contained by the conjugate diameters not be less than the angle contained by the lines AO, $O\Gamma$ inclined to the middle of the section, as Apollonius demonstrated in the penultimate proposition of Book II.[192] And if it will have made less than [that angle], the line $\Theta\Pi$ will come out greater than $\Theta\Xi$, and, accordingly the circle meant to meet the axis $A\Gamma$ cannot reach it.

We can also find these diameters, if [to and from] the given the sum of the squares on the axes, or rather the rectangle $A\Gamma\Delta$,[193] twice the rectangle contained by the conjugate [diameters] is added and subtracted, namely, [the rectangle] which is to the given rectangle contained by the axes of the ellipse as the ratio of ΘH to HT, or

---

[191] It is, of course, angle $A\Theta P$ that has been made equal to the given angle between the conjugates; but since $\Pi\Theta$ is perpendicular to $A\Theta$ and $P\Pi$ is perpendicular to $P\Theta$, it follows that angle $A\Theta P = \Theta\Pi P$.

[192] This is *Conics* II.52: "If a straight line touches an ellipse making an angle with the diameter drawn through the point of contact, it is not less than the angle adjacent to the one contained by the straight lines deflected at the middle of the section." Once again, the reader must keep in mind that the tangent at the vertex of a diameter is parallel to the conjugate diameter, so II.52 says that the angle *between the conjugate diameters* cannot be less than the said angle.

[193] Recall, $\Gamma\Delta$ is the latus rectum with respect to the axis $A\Gamma$.

as the radius to the side of the given angle.[194] Thus are obtained (by *Elem.* II.4, 7) the squares of the sum and difference of the diameters sought, *BK, ZH*.[195]

It ought to be observed here that in all these problems regarding the conjugate diameters of sections, we have investigated no diameters but these two, *BK,ZH*: the proposition can be equally applied to other diameters, diameters inclined to the axis at the same angle but to its other side.[196] One should note, furthermore, that in the preceding figures and demonstrations we set out the major axis as greater than its latus rectum: if the axis would be less than its latus rectum, nothing at all different or more difficult in the analysis or synthesis of the problem.

Scholium

The ancient geometers, and especially our Apollonius, had a method (mos) *for solving plane problems; afterwards, they developed this to the point* (rem eo deduxerant) *that [they could solve the problem of finding] the sides of a rectangle equal to a given rectangle, the difference or sum of whose sides was equal to a given line. This Euclid shows in* [propositions] *28 and 29 in the sixth book of the Elements where it is to be demonstrated how to apply a given parallelogram to a given line so that it exceeds or falls short by a parallelogram similar to a given one.*[197]

---

[194] Halley has in mind the angle within a circle as in the diagram below:

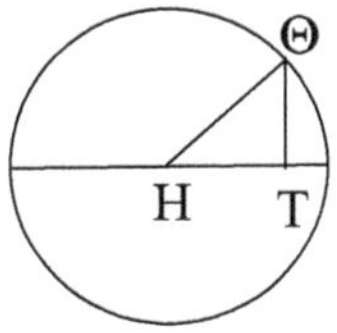

[195] The argument is this: The sum of the squares of the axes, which is equal to the rectangle $A\Gamma\Delta$, by *Conics* VII.12 is equal to the sum of the squares on conjugate diameters, sq.BK+sq.ZH. Since the ratio $\Theta H{:}HT$ is given (because the angle between the conjugate diameters is given) and since rect.BK,ZH:rect.Axes::$\Theta H{:}HT$, the rectangle contained by BK, ZH is given. But by *Elements* II.4,

$$\text{sq.}BK + \text{sq.}ZH + 2\text{rect.}BK,ZH = \text{sq.}(BK + ZH),$$

and, by *Elements* II.7,

$$\text{sq.}BK + \text{sq.}ZH - 2\text{rect.}BK,ZH = \text{sq.}(BK - ZH).$$

Therefore, $BK + ZH$ and $BK - ZH$ are given and, accordingly, BK and ZH.

[196] This observation seems to be only the trivial one that, with the exception of the axes, every pair of conjugate diameters containing a given angle, or a given area, or possessing any of the other properties treated in the previous theorems, has a matching pair "inclined to the other side" of the axis, that is, the pair obtained by reflecting the first in the minor or major axis.

[197] The exact enunciations of these propositions are as follows. *Elements* VI.28: "To a given straight line to apply a parallelogram equal to a given rectilineal figure and deficient by a parallelogrammic figure similar to a given one: thus the given rectilineal figure must not be greater than the parallelogram described on the half of the straight line and similar to the defect." *Elements* VI.29: "To a given straight line to apply a parallelogram equal to a given rectilineal figure and exceeding by a parallelogrammic figure similar to a given one." The cases which are of interest to Halley are those where the parallelogram to which the deficiency or excess is similar is a square; for in each of these cases a rectangle is found equal to a given rectangle and having sides whose sum or difference is equal to a given line (which happens to be the line to which the rectangles are

Special cases of this most general proposition are those of applying to a given line a rectangle or a square that exceeds or falls short by a square: geometers after Euclid needed results (effectionem postulant) *[following from] this. Since in practically all the subsequent problems these results—and none others from this point on will appear— truly will be of use, a brief list of them will be presented as simply as possible, and set forth in the manner of lemmas.*

Hence, to start, it is required to apply a given square to a given line exceeding by a square, that is, to find points Γ, Z, in a given line AB produced, so that the rectangles $A\Gamma B$, AZB are equal to the square on ΔE. Let AB be bisected at Δ, and let perpendicular ΔE, which is equal to the side of the square to be applied, be set up: let the lines $\Delta\Gamma$, ΔZ be made equal to the line AE, which has been joined. I say Γ, Z are the points sought.

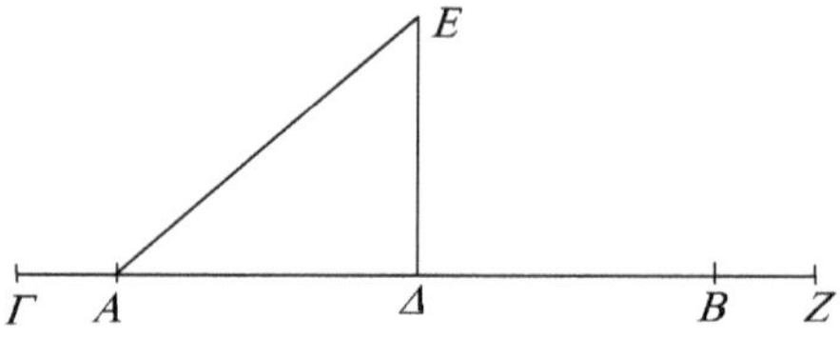

For the square on AE, that is, the square on $\Gamma\Delta$, is equal to the squares on $A\Delta$, ΔE taken together. But the square on $\Gamma\Delta$ (by Elem. II.6) is equal to the square on $A\Delta$ together with the rectangle $A\Gamma B$: therefore, the squares on $A\Delta$, ΔE are equal to the square on $A\Delta$ and the rectangle $A\Gamma B$; whence, having removed the common square on $A\Delta$, the square on ΔE is equal to the rectangle $A\Gamma B$, which was required. In the same way, the rectangle AZB can be shown to be equal to the same square on ΔE: whence it is clear that the lines $A\Gamma$, BZ are equal.

2nd *It is required to apply a given square to a given line falling short by a square, or to find in a given line AB, between A and B, points Γ, Z such that the rectangles $A\Gamma B$, AZB will be equal to the given square of some* [*line*] *ΔE.*[198] *Let AB be similarly bisected at Δ and let the perpendicular ΔE be the side of the given square; let an arc of a circle having center E and radius $A\Delta$ be described cutting line AB at points Γ, Z. I say that Γ, Z, are the points we are looking for.*

Indeed, the square on $E\Gamma$ is equal to the squares on ΔE, $\Gamma\Delta$ together, and the same square on $E\Gamma$, or rather $A\Delta$, is equal (by Elem. II.5) to the rectangle $A\Gamma B$ together with the square on $\Gamma\Delta$: removing the common square on $\Gamma\Delta$, the square on ΔE is, therefore, left equal to the rectangle AZB: whence $A\Gamma$ (or ZB) and AZ (or ΓB) are equal. And it is clear that DE must not be greater than half the given

applied!). Thus, suppose the given line is L and the given area is A: in the first case, we are looking for lengths a and b (b<L) such that rect.a,b=A and L-b=a (so that the deficiency is a square), that is, a+b=L; in the second case, we are looking for lengths a and b (b>L) such that rect.a,b and b-L=a, that is b-a=L.

198 It ought to be stressed that, in this case, these are two different solutions: (1) sq.ΔE=rect.$A\Gamma,\Gamma B$ =rect.$A\Gamma,(AB-A\Gamma)$=rect.$A\Gamma,AB$−sq.$A\Gamma$;
(2) sq.ΔE=rect.AZ,ZB=rect.$AZ,(AB-AZ)$ =rect.AZ,AB−sq.AZ.

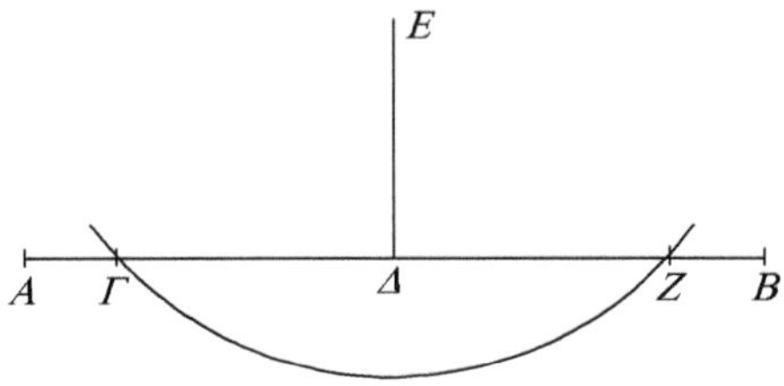

line AB, for otherwise the circle having center E and radius AD described would neither cut not touch AB; and with that the problem becomes impossible.

3rd A rectangle contained by given sides is to be applied to a given line exceeding by a square, that is, points H, Θ are to be found in the extended line AB so that the rectangle AΘB or AHB be equal to the rectangle contained by sides AΓ, BE. With the perpendiculars AΓ, BE set up at the ends of AB and on opposite sides, let ΓE be joined and bisected at Z; the circle described having center Z and radius ZΓ will pass through the desired points H, Θ.

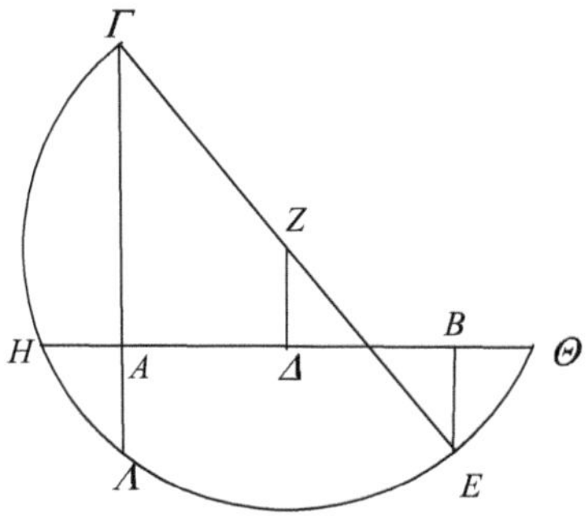

For drop a perpendicular ZΔ from the center Z; because of the equality of ΓZ, ZE, [the line] *AΔ will be equal to ΔB,*[199] *and, accordingly, AΛ will be equal to BE and AH to BΘ:*[200] *therefore, because of the circle, the rectangle ΓAΛ, that is, that contained by ΓZ, BE will equal (by Elem. III.35)*[201] *the rectangle HAΘ or AHB, and similarly to AΘB. Therefore the points H,Θ, which were sought, have been found.*

---

199 Let *BZ* be joined and extended to *M* on *ΓA*. Then, since the triangles *BZE* and *MZΓ* are congruent (*Elem.*, I.26), $BZ = ZM$. But since *ZΔ* is parallel to *ΓA* we then have immediately (by *Elem.*, VI.2) that $AΔ = ΔB$.

200 The word 'accordingly' (*proinde*) suggests that the equality of *AΛ*,*BE* and *HA*, *BΘ* follows from the equality of AΔ, ΔB. This is true for *HA*, *BΘ* (by *Elem.* III.14); that *AΛ* is equal to *BE*, however, follows more directly from the fact that *ΛE* is perpendicular to *ΓΛ* and, therefore, parallel to AB.

201 *Elements*, III.35: "If in a circle two straight line cut one another, the rectangle contained by the segments of the one is equal to the rectangle contained by the segments of the other."

4th A rectangle contained by given sides is to be applied to a given line deficient by a square; rather, points H, Θ are to be found on the line AB so that the rectangles AHB, AΘB be equal to the rectangle contained by the given [sides] AΓ, BE. Let perpendiculars AΓ, BE be set up at right angles [sic] and on the same side of AB; and let the joined [*line*] *ΓE be bisected at Z. Next, let a circle be described with center Z and radius ZΓ: it cuts the line AB (if the problem is possible) at the points sought, H, Θ.*

For let ZΔ be drawn perpendicularly to AB, and on account of ZE being bisected at Z, AΔ will be equal to ΔB and AΛ to BE [202]*: accordingly, the rectangle ΛAΓ, that is the rectangle contained by AΓ, BE will be equal (by Elem. III.36)*[203] *to rectangle HAΘ, that is, rectangle AHB or AΘB, and with that, points H, Θ are exhibited.*

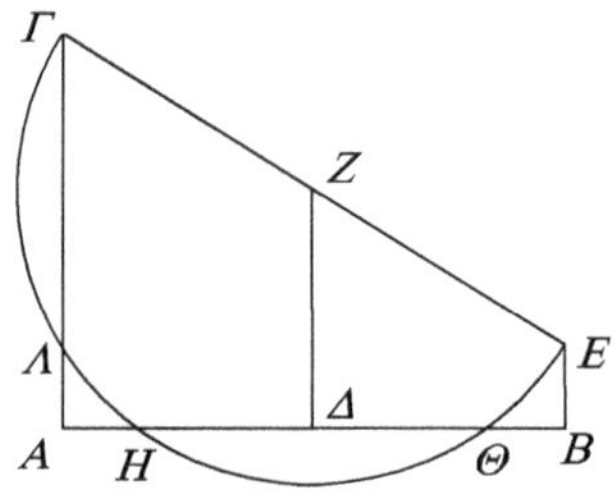

It is necessary, however, that the rectangle contained by AΓ, BE be not greater than the square on AΔ, since (by Elem. II.5) the square on AΔ is greater than the rectangle AHB by the square on HΔ; and, with that, the rectangle, AHB, that is, ΓAΛ, or, rather, that contained by the lines AΓ, BE, will not be greater than the square on AΔ or ΔB. If it were otherwise, the circle ΓΛE would not cut the line AB[204]*: whence, one must agree that the problem will be impossible.*

Proposition XIX (Problem)

Given in a hyperbola the axis and its latus rectum, find the diameter that has its latus rectum equal to a given line.

With the diagram of the hyperbola kept as in the previous [propositions], let the given straight line be ξ. It will be the case then (by VII.15)[205] that as the square on

---

202 See the note for the 3rd case.

203 *Elements* III.36: "If a point be taken outside a circle and from it there fall on the circle two straight lines, and if one of them cut the circle and the other touch it, the rectangle contained by the whole of the straight line which cuts the circle and the straight line intercepted on it outside between the point and the convex circumference will be equal to the square on the tangent."

204 Since, by *Elements* III.37 (which is the converse of *Elements* III.36), if the rectangle *ΓΛE* is equal to the square on *AΔ*, *AΔ* will be tangent to the circle.

205 *Conics* VII.15 formulates the proportion *alternando*: sq.*AΓ*:sq.ξ::rect.*NΓ*,*MΞ*: sq.*MN*.

the axis $A\Gamma$, or, rather, the rectangle contained by $N\Gamma$ and $\Gamma\Delta$ (the sum of the axis and the latus rectum), is to the rectangle contained by $N\Gamma$ and $M\Xi$, that is, as the sum of that axis and the latus rectum is to $M\Xi$, so is the square of the given latus rectum ξ to the square on MN. With that, if it be contrived that as the sum of the axis and its latus rectum is to the latus rectum ξ, so is ξ to some other line, say Φ, then the line Φ will be given, and the rectangle contained by $M\Xi$ and Φ will be equal to the square on MN[206]: whence $M\Xi$ will be to MN as MN is to Φ. Now, if it turns out that the axis be greater than its latus rectum, $M\Xi$ will be greater than MN, and, accordingly, MN will be greater than Φ : whence, by conversion, the ratio $M\Xi$ will be to ΞN just as MN to the excess by which MN is greater than Φ, and, *permutando*, $M\Xi$ will be to MN just as ΞN to the difference between MN and Φ. Again, by conversion, the ratio $M\Xi$ will be to ΞN just as ΞN is to the excess by which ΞN is greater than the difference of those same lines MN and Φ.

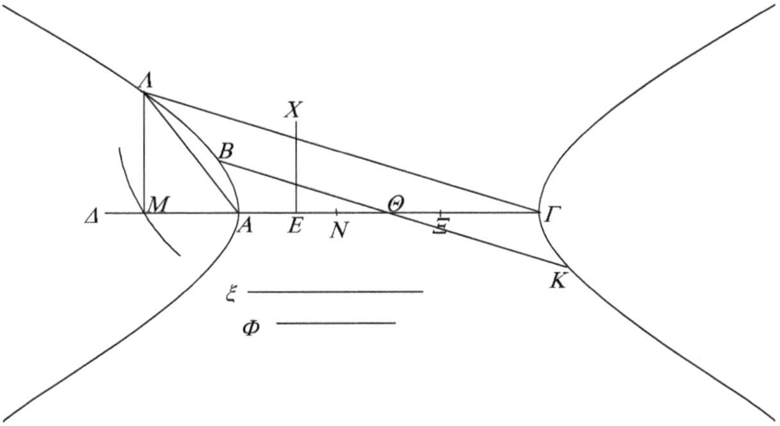

But MN is the excess by which $M\Xi$ is greater than ΞN; therefore, the excess by which ΞN is greater than the difference of MN and Φ is equal to the excess by which twice $N\Xi$ and Φ together is greater than $M\Xi$: therefore, as $M\Xi$ will be to $N\Xi$ so too will $N\Xi$ be to the excess by which twice $N\Xi$ and Φ together is greater than $M\Xi$: whence, the rectangle contained by $M\Xi$ and the excess by which twice $N\Xi$ and Φ together is greater than $M\Xi$ is equal to the square on $N\Xi$. But the square on $N\Xi$ is given; therefore the rectangle contained by $M\Xi$ and the aforementioned

[206] A classical argument would be as follows: starting with the proportion, $\Gamma\Delta$:$M\Xi$::rect.$\Gamma\Delta$,Φ: sq.MN, Φ is set as a common height to obtain rect.$\Gamma\Delta$,Φ:rect.$M\Xi$, Φ::rect.$\Gamma\Delta$,Φ: sq.MN. From this it follows that rect.$M\Xi$,Φ=sq.MN. However, the step in which a common height is taken (so that the ratio is changed from one between lines to one between figures) is generally stated explicitly in classical texts. The immediacy with which Halley states the conclusion leads me to suspect that, as we have seen before, Halley was using an algebraic formulation, that is, he writes $\Gamma\Delta/M\Xi=\Gamma\Delta\times\Phi/MN^2$, and then "cancels" $\Gamma\Delta$ to obtain $1/M\Xi=\Phi/MN^2$ or $M\Xi\times\Phi=MN^2$.

excess is given. Moreover, it is applied (*adjacet*) to a given line, namely, twice $N\Xi$ and Φ taken together, falling short by a square.[207] Therefore, $M\Xi$ is given, and the point M is given as well.

The synthesis, moreover, will be accomplished in this way. Let it be contrived that as the half-sum of the axis and its latus rectum is to half the latus rectum ξ, so too is the half latus rectum to half the line Φ, to which NE is made equal, and let EX be set up perpendicular to the axis and set equal to $N\Xi$, and with center X and radius ΞE let a portion of a circular arc be described cutting the axis at M, etc. The justification (*ratio*) for this is manifest from the analysis and from 2nd lemma of the Scholium.[208]

If in fact the axis of the hyperbola turns out to be less than the its latus rectum, $M\Xi$ will be less than MN. But if, as before, $M\Xi$ is to MN as MN is to Φ, Φ will be greater than MN; whence, by conversion, the ratio $M\Xi$ to $N\Xi$ will be just as MN is to the excess by which Φ is greater than MN, and *permutando*, $M\Xi$ will be to MN just as $N\Xi$ is to the excess of Φ over MN. With that, again, by conversion, the ratio $M\Xi$ to $N\Xi$ will be as $N\Xi$ is to the excess by which the difference between Φ and twice $N\Xi$ is greater than $M\Xi$. Therefore the rectangle contained by $M\Xi$ and the excess by which $M\Xi$ is surpassed by the difference which is between Φ and twice $N\Xi$ will be equal to the square on $N\Xi$. But the square on $N\Xi$ is given: therefore, the rectangle contained by $M\Xi$ and the excess mentioned is given. Moreover, this given rectangle is applied to a given line, namely, the excess by with Φ is greater than twice $N\Xi$, falling short by a square. Therefore, $M\Xi$ is given, and the point M is given as well.

The synthesis is hardly different, except that in this case the point N is further from the vertex than Ξ : therefore, let it be contrived that as the half-sum of the axis and its latus rectum is to half the given latus rectum, so too is that same half given latus rectum to a third proportional, to which NE is set equal[209]; and with EX set up perpendicular to the axis, let EX be made equal to $N\Xi$; and with center X and radius ΞE let a circle be drawn cutting the axis at the point M sought, or in points M, μ, depending on how many times it can come about[210]; for NE is equal to half Φ, and with that ΞE is equal to half [the line] to which the rectangle equal to the square on $N\Xi$ is applied, namely, half the excess by which Φ is greater than twice $N\Xi$.

Diorism.[211] In the first case, where the axis is greater than its latus rectum, it is clear (by VII.33)[212] that the latus rectum will be less than the latus rectum of any

---

[207] In other words, the problem of finding $M\Xi$ has been reduced to a typical problem of "application of areas," the subject of the Scholium just prior to proposition XIX.

[208] The application of the 2nd lemma of the Scholium is clear by comparing the diagrams: E here corresponds to Δ in the Scholium, X to E, M to Γ, Ξ to B. A point corresponding to A is unnecessary since E has been constructed so that $E\Xi$ is equal to $\frac{1}{2}(2N\Xi+\Phi)$, that is, half the line to which the square on $N\Xi$ is to be applied.

[209] That is, (axis + latus rectum)/2: $\xi/2 :: \xi/2 : \Phi/2 (= NE)$.

[210] In other words, whether the arc cuts the axis once or twice inside the hyperbola.

[211] Always written in Greek in the text: thus, διορισμὸς and διορισμοὶ.

[212] *Conics* VII.33 states: "If there is a hyperbola, and the transverse diameter of the figure constructed on the axis is not less than its *latus rectum*, then the *latus rectum* of the figure constructed

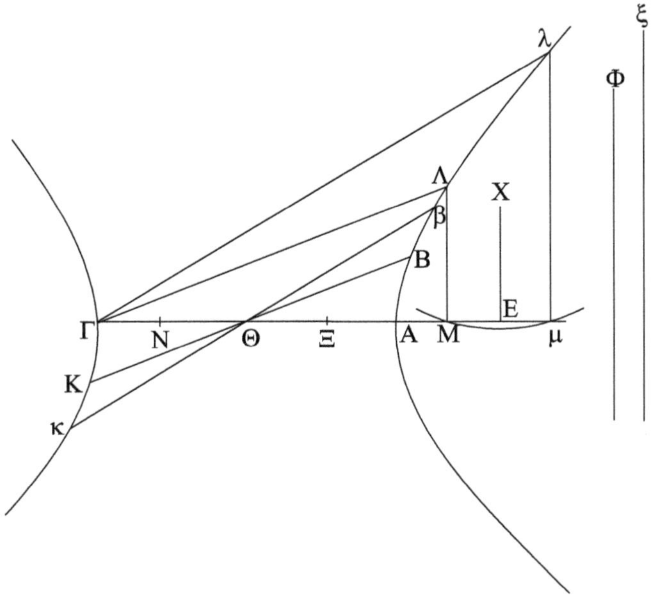

other diameter; and with that, the proposed latus rectum ξ must be greater than the latus rectum of the axis; and ξ will be large to the degree the diameter sought is farther from the axis of the section. And furthermore, in the other case, if the axis happens to be less than its latus rectum, though not less than half of it, the same will hold (by VII.34).[213] But if, in fact, the latus rectum happens to be greater than twice the axis, then $N\Xi$ will be greater than ΞA[214]: and if it be contrived that ΞM is equal to $N\Xi$, and a perpendicular *MX*, or *EX*, equal to $N\Xi$, is set up, then (by VII.35)[215] there will be in the section a diameter, *BK*, whose latus rectum will be twice the diameter[216] [and will be] a minimum among all the latera recta, since [in that case]

---

on the axis is less than the *latus rectum* of [any of] the figures constructed on the other diameters of the section, and the *latus rectum* of [any of] the figures constructed on diameters closer to the axis is less than the *latus rectum* of the figures constructed on [diameters] farther from the axis."

[213] *Conics* VII.34 demonstrates that the claim of VII.33 holds also when the length of the axis is less than that of its latus rectum, but not less than half of it.

[214] For since latus rectum:axis::$AN : A\Xi$, (latus rectum−axis): axis :: $(AN - A\Xi) : A\Xi :: N\Xi : A\Xi$. But since latus rectum> 2×axis, (latus rectum−axis)>axis>axis:axis, therefore, also $N\Xi : A\Xi > A\Xi : A\Xi$, so that $N\Xi > A\Xi$.

[215] *Conics* VII.35 states (continuing from VII.33 and VII.34): "Furthermore, we make $A\Gamma$ [the axis (MNF)] less than half the *latus rectum* of the figure of the section constructed on it: then I say that there are two diameters, [one] on either side of this axis, such that the *latus rectum* of the figure constructed on each of them is twice that [diameter]; and that [*latus rectum*] is less than the *latus rectum* of the figure constructed on any other of the diameters on that side [of the axis]; and the *latus rectum* of figures constructed on the diameters closer to those two diameters is less than the *latus rectum* of a figure constructed on a [diameter] farther [from them]."

[216] By *Conics* VII.6, *BK*: latus rectum (BK) :: $\Xi M : MN$. But ΞM has been contrived to be equal to $N\Xi$, or $\frac{1}{2}MN$; therefore, *BK* must be a diameter which is half its latus rectum.

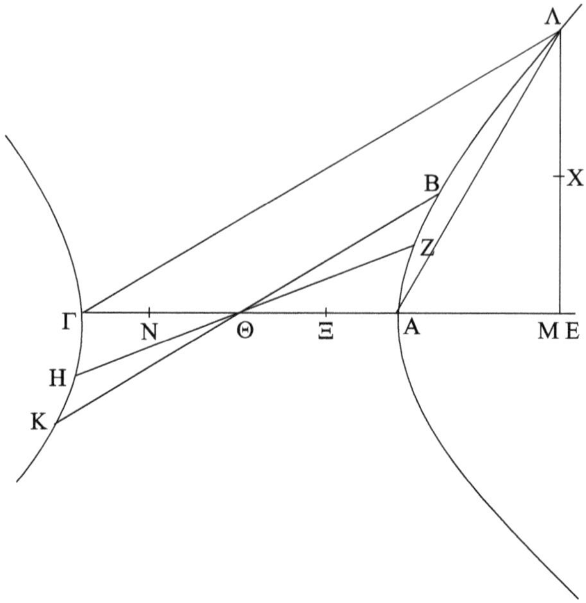

the points then M and E coincide[217] and a circle whose center is X and whose radius is $M\Xi$ is tangent to the axis at point M, because ΞE is equal to EX. But the diameter BK, by what has been shown in the sixth proposition of this [work],[218] is a mean proportional between $M\Xi$, or $N\Xi$, and the sum of the axis and its latus rectum; and with that, the rectangle contained by $N\Xi$ and the sum of the axis and its latus rectum is equal to the square on BK. But the sum of the axis and latus rectum is to the difference of the same as the axis $A\Gamma$ is to $N\Xi$; whence the rectangle contained by $N\Xi$ and the sum of the axis and its latus rectum is equal to the rectangle contained by the axis and the excess of the latus rectum over the axis: therefore, the square on BK is equal to the rectangle contained by the axis and the difference of the axis and latus rectum, that is, the difference between the figure of the axis and the square of the axis: therefore, BK will be a mean proportional between the axis and the difference of the axis and the latus rectum; and the minimum latus rectum of the hyperbola will be twice BK.

So, if the proposed latus rectum happens to be less than twice the mean proportional between the axis and the difference of the axis and its latus rectum, that is, if the square of [the proposed latus rectum] happens to be less than four times the

[217] Recall NE is set equal to $\Phi/2$ and M is determined by the relation sq.$N\Xi$=rect.$M\Xi,(\Phi - 2N\Xi)$−sq.$M\Xi$ = rect. $M\Xi,(2NE - 2N\Xi)$−sq.$M\Xi$. But $N\Xi = \Xi M$, which from the latter relation, implies that $NE = 2M\Xi$. Thus, since $NE = N\Xi + \Xi E = M\Xi + \Xi E$, we have $\Xi E = \Xi M$, that is, E and M coincide.

[218] This of course is a mistake. In the sixth proposition, Halley shows that BK is a mean proportional between $M\Xi$ and $\Gamma\Delta$ in the case of an ellipse; it is in the fifth proposition he demonstrates the corresponding fact for the hyperbola.

excess by which the rectangle contained by the axis and its latus rectum is greater than the square of the axis, then the problem will be impossible. If this [latus rectum] turns out to be greater [than the aforementioned mean proportional], but less than the latus rectum of the axis, two diameters will be found on each side of the axis to which the given latus rectum will correspond (*competat*): if [the proposed latus rectum] is equal to the latus rectum of the axis, then, beside the axis itself, one [diameter] will be found on either side [of the axis], so that there will three diameters in all meeting the requirement. If the proposed latus rectum happens to be greater than the latus rectum of the axis, there will be only one diameter on either side of the axis that can satisfy the problem. A *maximum*, however, is not given.[219]

I say, in addition, that just as the latus rectum of diameter BK is twice BK, so, in every case where one has two diameters on either side of the axis whose latera recta are equal, the sum of the same is equal to their common latus rectum.

For (by VII.29) the difference of the square of some diameter ZH and the figure on ZH is equal to the difference of the square of the axis $A\Gamma$ and the figure of it, that is, it is equal to the rectangle contained by the axis and the difference between the axis and its latus rectum: therefore, the rectangle contained by ZH and the excess by which its latus rectum is greater than ZH is given. But it is applied to a given line, the proposed latus rectum, and deficient by a square: accordingly, the latus rectum will be equal to both that of ZH and of another diameter which has the same latus rectum as ZH.[220]

Corollary. Hence, it is clear that another diameter exist, which has the same latus rectum as the axis $A\Gamma$, is equal to the excess by which that latus rectum is greater than the axis.

Proposition XX (Problem)

Given in an ellipse the axis and its latus rectum, find the diameter that has its latus rectum equal to a given line.

With the diagram of the ellipse kept as before, let the given straight line ξ. It is necessary, then, to find that diameter of this ellipse which has its latus rectum equal to ξ. By VII.15, it is shown that the square on $A\Gamma$, or the rectangle contained by $N\Gamma$ and $\Gamma\delta$ (the difference of the axis and its latus rectum), is to the rectangle contained by $N\Gamma$, $M\Xi$ just as the square of the latus rectum ξ is to the square on MN. Therefore, if it is contrived that as the difference of the axis and its latus rectum is to ξ, so too is ξ to some other [line], say, Φ , the line Φ will be given, and the rectangle contained by $M\Xi$ and Φ will be equal to the square on MN.[221] Therefore, the analogy (ἀνάλογον)[222] is that $M\Xi$ will be to MN just as MN is to Φ, and,

---

[219] In other words, for the problem to have a solution, ξ can be as large as one pleases.

[220] Halley's exposition is not entirely clear. The argument from VII.29 can be formulated as follows. First, observe that by VII.35 there can be unequal diameters having the same latus rectum. Suppose AB and CD are unequal diameters with the same latus rectum L, then, by VII.29, $\text{rect.}AB,L-\text{sq.}AB=\text{rect.}CD,L-\text{sq.}CD$ or $\text{rect.}(AB-CD),L=\text{sq.}AB-\text{sq.}CD$, from which it follows that $L=AB+CD$.

[221] The argument is as in VIII.19.

[222] See the note on the word "analogy" in proposition XI.

componendo, ΞN will be to MN just as MN and Φ together are to Φ; whence the rectangle contained by ΞN and Φ will be equal to the square on MN together with the rectangle contained by MN and Φ. But the rectangle contained by ΞN, Φ is given; therefore, the rectangle contained by MN and MN and Φ together is given. Therefore, the given rectangle contained by ΞN and Φ is applied (*adjacet*) to the given line Φ and exceeding by a square; whence, the line MN is given. And, on account of the given point N, the point M will also be given.

The synthesis, therefore, is clear: for if ΞN is produced to Π, and $N\Pi$ is made equal to Φ, or $N\Pi$ [is made so that it be] to ξ just as ξ to $\Gamma\delta$, the difference of the axis and its latus rectum; and to $N\Pi$ is applied (*applicetur*) the rectangle equal to that contained by ΞN and $N\Pi$, exceeding by a square, which is the rectangle $NM\Pi$,[223] the point M will be found (by the 3rd Lemma of the Scholium), whence the position of BK, which satisfies the problem, will be obtained.

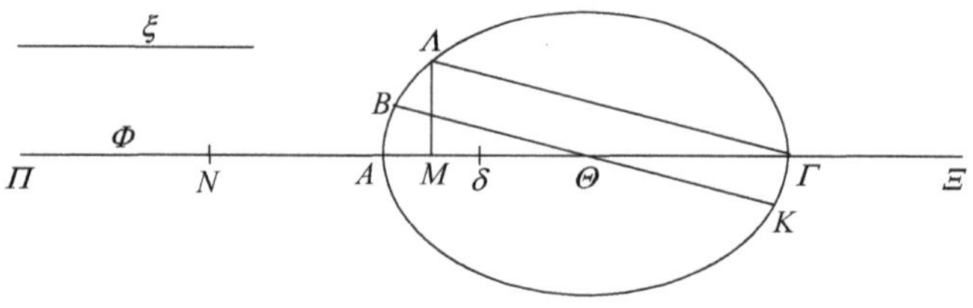

This problem for the ellipse can be solved differently, in the way used for the hyperbola. It comes from a slightly more involved construction: for when $M\Xi$ is to MN just as MN is to Φ, *componendo*, $M\Xi$ will be to ΞN just as MN to MN andΦ together, and, *permutando*, $M\Xi$ will be to MN just as ΞN is to MN and Φ together; and again, *componendo*, $M\Xi$ will be to ΞN just as ΞN is to MN, ΞN and Φ taken together, or to the excess by with Φ and twice ΞN is greater than $M\Xi$.[224] Therefore, the square on ΞN is equal to the rectangle contained by $M\Xi$ and the excess by which Φ and twice ΞN is greater than $M\Xi$. That rectangle, indeed, is given, on account of the given [line] $N\Xi$; it is, in fact, applied to the given line, namely, that which is equal to Φ and twice $N\Xi$ together, and deficient by a square: therefore, the line $M\Xi$ is given, and, on account of the given point Ξ, the point M is also given.

Such a construction is accomplished thus. Let it be contrived that as the difference of the axis and its latus rectum is to ξ, so too is half ξ to half Φ, to which NE is made equal; and perpendicular EX having been set up, let EX be made equal to $N\Xi$, and with center X and radius ΞE, let a circle be described cutting the axis at point M. This being found, the rest will be completed as before.

And the *diorism* for this problem is clear. For if the proposed latus rectum happens to be less than the latus rectum of the major axis, or greater than the latus rectum of the minor axis, the problem will be impossible, the point M falling between E and A in the former case, and beyond the vertex Γ in the latter.

---

[223] The rectangle $NM\Pi$ is that contained by NM and $M\Pi$, or MN and $MN+N\Pi$ $(=MN+\Phi)$.

[224] Since $MN = N\Xi - M\Xi$, we have $\Xi N + MN + \Phi = \Xi N + N\Xi - M\Xi + \Phi = 2N\Xi + \Phi - M\Xi$.

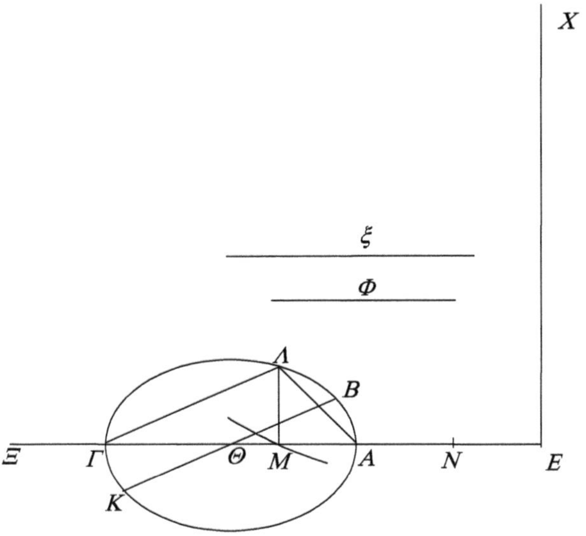

Proposition XXI (Problem)

Given the axis of a hyperbola and the latus rectum of the axis, find a diameter which has to its latus rectum a given ratio.

With the description kept as before, let the given ratio be as Σ to T, and let BK be set as the diameter sought; having dropped a perpendicular ΛM, [it follows that] as Σ is to T, or BK to its latus rectum, so too (by VII.6) will $M\Xi$ be to MN. Therefore, the ratio of $M\Xi$ to MN will be given: and, *dividendo*,[225] the ratio $N\Xi$ to ΞM will be given, which of course is the same as the ratio of the difference of the terms Σ and T to the term Σ, corresponding to the diameter (*diametro analogum*). And on account of the given $N\Xi$, [the line] ΞM is given as well, whence also the point M is given.

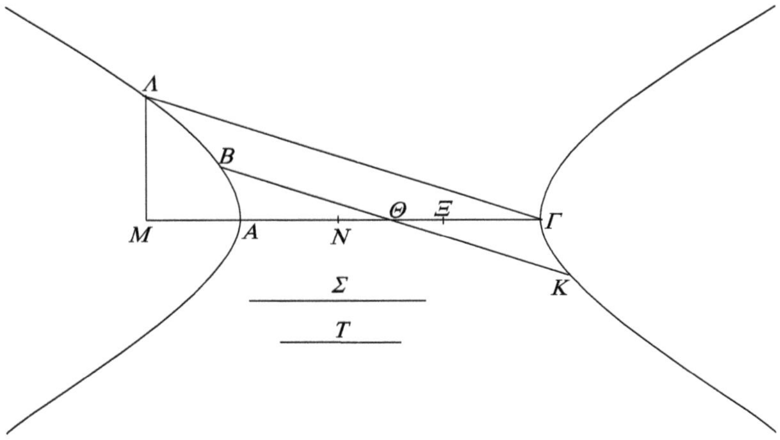

[225] This is essentially the same as the more conventional *separando*.

Therefore, if it be contrived that as the difference of the terms is to the term corresponding to the diameter so too is $N\Xi$ to ΞM; the point sought, M, will be obtained, and, whence, all the rest will follow.

But the proposed ratio cannot be greater than the ratio of the axis to its latus rectum, if the axis happens to be greater than its latus rectum; nor less than that same ratio, if the axis happens to be less.[226]

Proposition XXII (Problem)

Similarly, given the axis of an ellipse and latus rectum of the axis, a diameter is to be found which has a given ratio to its latus rectum.

With the diagram of the ellipse kept as before, let the ratio be as Σ to T. Consider it done, and let BK be the diameter we are seeking: therefore (by VII.7), as Σ is to T, that is as BK to its latus rectum, so too will $M\Xi$ be to MN. Therefore, the ratio $M\Xi$ to MN is given, and, *componendo*, the ratio $N\Xi$ to ΞM is given; for the same is also the ratio of the sum of the terms Σ, T to the term Σ which corresponds (*respondet*) to the diameter. But ΞN is given; and with that $M\Xi$ is also given, and the point M is given.

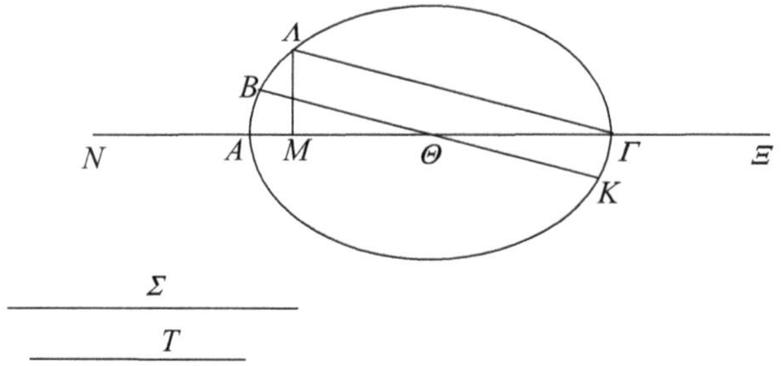

If, therefore, it be contrived that as the sum of the terms Σ, T to the term Σ be as $N\Xi$ to ΞM, we shall have the point M, and, with the help of that, [we shall have] the diameter sought both in magnitude and position, as we showed in the sixth proposition of this [work].

But the proposed ratio cannot be greater than the ratio of the major axis to its latus rectum; nor less than the ratio of the minor axis to its latus rectum; that is not less than the ratio of the latus rectum of the major axis to that same axis.[227]

Proposition XXIII (Problem)

Given the axis and latus rectum of a hyperbola, it is required to find the position and magnitude of a diameter which differs from its latus rectum by a given difference.

[226] These *diorisms* follow from *Conics* VII.21 and 22, respectively.

[227] Again, these follow from the corollaries after *Conics* VII.31.

With the figure of the hyperbola kept [as before], suppose [the problem] is done; and let the diameter sought be BK, [let] $\Gamma\Lambda$ be parallel to the same, and let the perpendicular ΛM be dropped. Then (by VII.16)[228] as the square on $A\Gamma$, or the rectangle contained by $N\Gamma$, $\Gamma\Delta$, will be to the rectangle on $N\Gamma$, $M\Xi$ (that is, as $\Gamma\Delta$ will be to $M\Xi$), so too will the square of the difference of the diameter BK and its latus rectum be to the square on ΞN. But the ratio of this difference to the square on ΞN is given, on account of the very givens [of the problem]:[229] whence the ratio of $\Gamma\Delta$ to $M\Xi$ will be given; and on account of the given line $\Gamma\Delta$,[230] the line $M\Xi$ is also given, and, with that, the point M [is given].

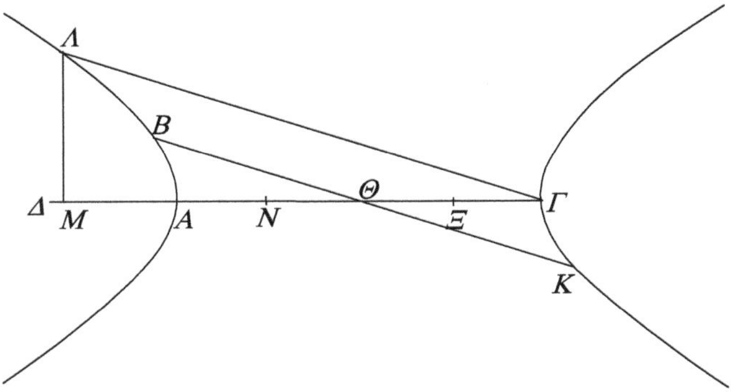

The synthesis, moreover, will be accomplished in this way. Let it be contrived that as the square of the given difference is to the square on ΞN, so too is the sum of axis and its latus rectum to $M\Xi$: but the point M being found, the rest will be completed in the way said above.

And it is understood (*constabit*) that the given difference should be less than the difference between the axis and its latus rectum, from those things demonstrated in VII.6, and from the same construction[231]: for all the lines $M\Xi$ are reciprocally related to (*reciproce ut…*) the squares of the differences between the diameters and their latera recta. And, with that, they continually increase as long as those differences decrease, so they never can reach a *minimum*.[232]

---

228 *Conics* VII.16 formulates the proportion *alternando*, sq.$A\Gamma$:sq.($BK-$latus rectum(BK)):: rect.$N\Gamma$,$M\Xi$:sq.($MN - M\Xi$). There is no general enunciation for VII.16.

229 ΞN is given of course because the points Ξ and N are determined by the ratio of the axis to its latus rectum, which is given.

230 That is, because $\Gamma\Delta$ is given.

231 No doubt VII.36, and not VII.6, was meant. For *Conics* VII.36 demonstrates that the difference between the axis and its latus rectum is greater than that between any other diameter and its latus rectum.

232 By "reciprocally," Halley means something like "inversely." Indeed, $M\Xi$:$\Gamma\Delta$::ΞN:sq.(BK-lr.BK), while $\Gamma\Delta$ and ΞN are fixed. Therefore, as Halley says, $M\Xi$ grows continually as BK-lr.BK decreases; hence, while there is an upper bound for BK-lr.BK, namely, the difference between the axis and its latus rectum, there is no lower bound, or, as Halley puts it, the differences "never can reach a minimum."

We will obtain the magnitudes of the diameters, however, by a different argument: for since the difference between the square of any diameter and its figure (by VII.29) is always equal to the difference between the square of the axis and its figure, the analogy (ἀνάλογον) is that as the given difference between the diameter sought and its latus rectum is to the difference between the axis and its latus rectum, so to will the axis of the hyperbola be to the diameter sought. Whence it is manifest that those differences are everywhere reciprocally proportional to the corresponding diameters (*suis diametris*).[233]

Proposition XXIV (Problem)

Moreover, in an ellipse, given the axis and its latus rectum, it is required to find a diameter of the section, which differs from its latus rectum by a given difference.

With the previous description for the ellipse kept [as before], consider [the problem] done; and let BK be the diameter which we are seeking. By VII.16, it is demonstrated that the square on $A\Gamma$, or the rectangle contained by $N\Gamma$, $\Gamma\delta$, is to the rectangle contained by $N\Gamma$, $M\Xi$, that is, $\Gamma\delta$ to $M\Xi$, just as the square on half the difference proposed between BK and its latus rectum and the square on ΘM.[234] Therefore, if it be contrived that as $\Gamma\delta$, or the difference between the axis and its latus rectum, is to half the proposed difference, so too is the same half-difference to another [line], say, to ψ; the line ψ will be given; and the rectangle contained by ψ and $M\Xi$ will be equal to the square on ΘM.[235] Accordingly, the analogy (ἀνάλογον) is that $M\Xi$ will be to ΘM just as ΘM is to ψ; and, *dividendo*, or *componendo*,[236] $\Xi\Theta$ will be to ΘM just as the difference, or sum, of ΘM and ψ is to ψ. With that, the rectangle contained by $\Xi\Theta$ and ψ will be equal to the rectangle contained by ΘM and the sum, or difference, of ΘM and ψ, which rectangle is, accordingly, given. Therefore, a rectangle equal to the rectangle contained by $\Xi\Theta$ and ψ is applied to the given line ψ and exceeding by a square; because ψ is given by the difference of the lines, therefore (*by Lemma 3 of our Scholium*), the line ΘM is given; and, on account of the given Θ, the point M is also given.

Whence, a synthesis of the following kind arises (*oritur*). Let it be contrived that as the difference of the axis and its latus rectum is to the half-difference proposed for the diameter and its latus rectum, so too is that same half-difference to a fourth [proportional], namely, to ψ, to which ΘE along the axis is set equal: and producing

---

[233] By VII.29, $\text{sq.}BK - \text{rect.lr.}BK, BK = \text{sq.Ax} - \text{rect.lr.Ax,Ax}$ or $\text{rect.}BK,(BK - \text{lr.}BK) = \text{rect.Ax,}(\text{Ax} - \text{lr.Ax})$. Therefore, if D is the given difference, $BK - \text{lr.}BK$, and D_{Ax} is the difference between the axis and its latus rectum, $\text{Ax} - \text{lr.Ax}$, we have the "reciprocal" relation: $BK : \text{Ax} :: D_{\text{Ax}} : D$. Since Ax, D_{Ax}, and D are all given, the magnitude of BK is given.

[234] By *Conics* VII.16, we have $\Gamma\delta{:}M\Xi{::}\text{sq.}1/2(BK\text{-lr.}BK){:}\text{sq.}1/2(M\Xi - MN)$. But $M\Xi = M\Theta + \Theta\Xi = M\Theta + \Theta N$ and $MN = \Theta N - M\Theta$. Therefore, $1/2(M\Xi - MN) = 1/2(M\Theta + \Theta N - (\Theta N - M\Theta)) = M\Theta$. So, $\Gamma\delta{:}M\Xi{::}\text{sq.}1/2(BK\text{-lr.}BK){:}\text{sq.}M\Theta$.

[235] For $\text{rect.}\Gamma\delta,\psi : \text{rect.}M\Xi,\psi :: \text{sq.}1/2(BK\text{-lr.}BK) : \text{sq.}M\Theta$. But ψ has been defined so that $\Gamma\delta{:}1/2(BK\text{-lr.}BK){::}1/2(BK\text{-lr.}BK){:}\psi$, that is, so that $\text{rect.}\Gamma\delta,\psi{=}\text{sq.}1/2(BK\text{-lr.}BK)$. Thus, $\text{sq.}1/2(BK\text{-lr.}BK){:}\text{rect.}M\Xi,\psi{::}\text{rect.}\Gamma\delta,\psi{:}\text{rect.}M\Xi,\psi{::}\text{sq.}1/2(BK\text{-lr.}BK) : \text{sq.}M\Theta$, from which it follows that $\text{rect.}\Gamma\delta,\psi{=}\text{sq.}M\Theta$.

[236] Whether it is *dividendo* or *componendo* depends on whether M happens to fall between Θ and N or between Θ and Ξ.

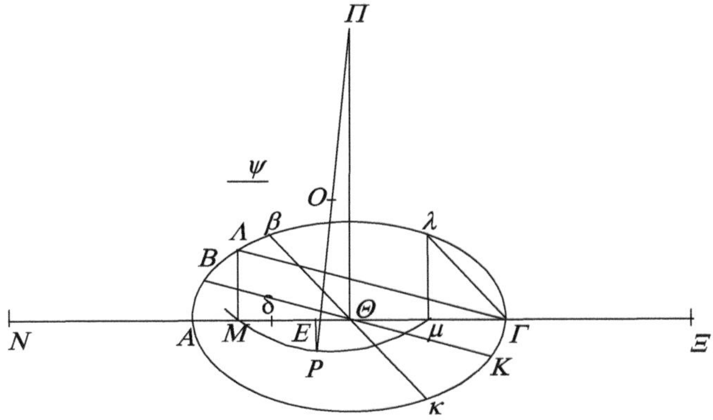

the minor axis to Π such that $\Theta\Pi$ be equal to $\Theta\Xi$, let a parallel PE be drawn equal to $E\Theta$: and let ΠP be joined, which is bisected at O, and the circular arc $MP\mu$ drawn with center O and radius PO will, if the problem be possible, cut the axis at points M, μ, or at μ alone, and from here the rest is plain [*uti in sequentibus patebit*].

Another and slightly simpler solution for the problem, moreover, may be obtained. For since $M\Xi$ is to ΘM as ΘM to ψ, by conversion, the ratio $M\Xi$ to $\Xi\Theta$ is just as ΘM to the excess by which ΘM is greater than ψ, and, *permutando*, $M\Xi$ will be to ΘM as $\Xi\Theta$ is to the just mentioned excess: whence, again by conversion of the ratio, $M\Xi$ will be to $\Xi\Theta$ just as $\Xi\Theta$ is to the excess by which $\Xi\Theta$ and ψ taken together are greater than ΘM, that is, to the excess by which $N\Xi$ and ψ together are greater than $M\Xi$.[237] Therefore, the square on $\Xi\Theta$ is equal to the rectangle contained by $M\Xi$ and excess by which $N\Xi$ and ψ together are greater than $M\Xi$, which rectangle is given, on account of the given square on $N\Xi$.[238] But that rectangle is applied to a line equal to $N\Xi$ and ψ together deficient by a square: and, accordingly (*by Lemma 2 of our Scholium*) the line ΞM is given, and, on account of the given point Ξ, M is given.

The synthesis, therefore, will be accomplished in this way. Let ΘE be made equal to half ψ and placed from Θ towards N; and let EY be set up perpendicularly, and set EY equal to $\Theta\Xi$. Next, with center Y and radius ΞE let an arc of a circle $M\Sigma\mu$ cutting the axis at points M, μ be described, from which is obtained the position of a diameter, on either side, which [differs] from its latus rectum by the proposed difference. But in the one case the diameter exceeds the latus rectum, while in the other the latus rectum is greater than the diameter by the same difference.

---

237 The argument is similar to that in Proposition XX: $(\Theta\Xi + \psi) - \Theta M = NM + \psi = N\Xi - M\Xi + \psi = (N\Xi + \psi) - M\Xi$.

238 Why Halley puts it this way is unclear: indeed, sq.$N\Xi$ and sq.$\Xi\Theta$ are equally given since $N\Xi = 2\Xi\Theta$ or sq.$N\Xi$=4sq.$\Xi\Theta$.

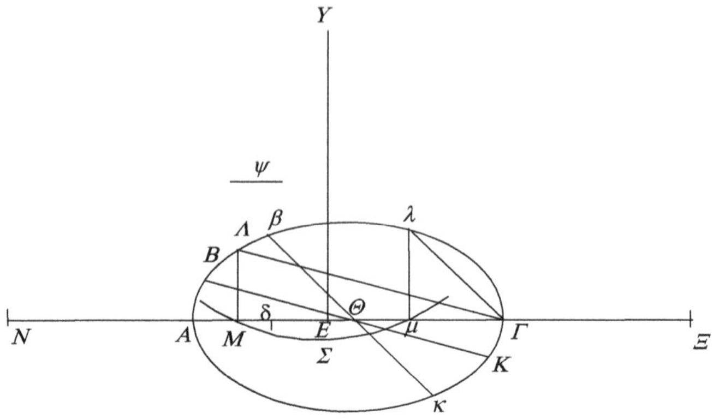

For this problem, moreover, one appeals to VII.37 for the *diorisms*.[239] For if the proposed difference turns out to be greater than that by which the major axis is greater than its latus rectum, the point M will fall outside the axis, beyond the vertex A. And if the difference turns out to be greater than that between the minor axis and its latus rectum, the point μ will also fall beyond the vertex Γ: whence the problem will be altogether impossible. If indeed [the difference proposed] turns out to be less than [the difference by which the minor axis is greater than its latus rectum] but greater than that which exists between the major axis and its latus rectum, then the problem will be solved by two diameters adjacent to the minor axis on both sides. If the proposed difference turns out to be less than that between the major axis and its latus rectum, M and μ will each fall in the axis $A\Gamma$, and there will be

239 *Conics* VII.37 states: "[In] every ellipse, for the figures of the section constructed on the diameters greater than their [corresponding] *latera recta*, the difference between the two sides of the figure constructed on the major axis is greater than the difference between the [two] sides of [any of] the figures constructed on the remaining [diameters]; and the difference between the [two] sides of those [figures] constructed on [diameters] closer to the major axis is greater than the difference between the [two] sides of those [figures] constructed on [diameters] farther [from the major axis].

"But in the case when the diameters on which the figures are constructed are less than the [corresponding] *lateral recta*, the difference between the two sides of the figure constructed on the minor axis is greater than the difference between the [two] sides of the others of these figures; and the difference between the [two] sides of those [figures] constructed on [diameters] farther from it.

"And the difference between the two sides of the figure constructed on the minor [this is Halley's correction—Toomer notes the error in the Arabic text. (MNF)] axis is greater than the difference between the two sides of the figure constructed on the major axis [again Halley's correction. (MNF)]."

To understand this proposition—and, therefore, also Halley's *diorisms*—it is important to keep in mind that in every ellipse there is a diameter (and of course its equal conjugate) which is equal to its latus rectum. Let us call this diameter E; and so E=lr.E. For all diameters D between E and the major axis, D>lr.D; for all diameters d between E and the minor axis, d<lr.d. *Conics* VII.37 tells us then that: (1) for those diameters D between the major axis and E, Maj.Ax-lr.Maj.Ax>D-lr.D>D′-lr.D′, where D′ lies farther from the major axis than D; (2) for those diameters d between E and the minor axis, lr.min.ax-min.ax>lr.d-d>lr.d′-d′, where d′ lies farther from the minor axis than d; (3) lr.min.ax-min.ax.>Maj.Ax-lr.Maj.Ax.

altogether four different diameters whose differences from their latera recta are all equal to one another and to the given difference. Moreover, there is no *minimum* for the difference, but in equal conjugate diameters it [the difference] will vanish, M and μ coinciding [in that case] at the center Θ.

Corollary 1. It can be also demonstrated without further ado that the difference of the two diameters exhibiting the same proposed [difference] is equal to half the given difference between the diameters and their latera recta.[240] With that, if one of these diameters is given with its latus rectum, you will easily find the other. For the sum and difference of the givens (of the diameter and the half-difference) will be equal to the other diameter in the one case and to the latus rectum in the other, which were sought.[241]

Corollary 2. Whence, twice any diameter will equal another diameter together with its latus rectum, whose difference be equal to the difference between the given diameter and its latus rectum.[242]

Corollary 3. By the same argument, it is plain that a diameter of an ellipse whose conjugate is equal to it will be the mean proportional between any two diameters of the section for which the one is greater than its latus rectum by the same excess by which the latus rectum of the other is greater than the diameter.

Proposition XXV (Problem)

Given the axis in a hyperbola and its latus rectum, it is required to find the position of that diameter which together with its latus rectum comes to a given sum.

With same things set out as in the last diagram from the hyperbola, consider what we want has been done; and let BK be that diameter which with its latus rectum makes the given sum. By VII.17, the square on $A\Gamma$, or the rectangle $N\Gamma\Delta$, that is, that contained by $N\Gamma$ and the axis and its latus rectum together, is to the rectangle contained by $N\Gamma$ and $M\Xi$, just as the square of the sum of any diameter and its latus rectum is to the square of the line made up of NM and $M\Xi$ taken together.[243] Therefore, as the sum of the axis and its latus rectum is to $M\Xi$, so too will the square on BK and its latus rectum together be to the square on NM, $M\Xi$ together, or to four

[240] This is less transparent than Halley makes it seem, and, as is often the case, I suspect he was working algebraically; I will do the same. Let the given difference be K, and let the two diameters be D and d. Hence, we have the following relations: (1) D-lr.D=lr.d-d=K (the defining relation), (2) D+d=lr.D+lr.d (by subtraction), and (3) D-d+lr.d-lr.D=2K or lr.d-lr.D=2K-(D-d) (by addition). Squaring both sides of the relation (1), we have: D^2-2Dlr.D+lr.D^2=d^2-2dlr.d+lr.d^2 or D^2- d^2 =2(Dlr.D-dlr.d)+lr.d^2 – lr.D^2. But, by *Conics* VII.30, we have Dlr.D-dlr.d= d^2- D^2, while, by the relations (2) and (3), we have lr.d^2-lr.D^2=(2K-(D-d))(D+d). Thus, substituting these into the relation D^2- d^2 =2(Dlr.D-dlr.d)+lr.d^2 – lr.D^2, we have D^2- d^2=2(d^2- D^2)+2K(D+d)-(D^2- d^2)=2K(D+d)-3(D^2- d^2) or 4(D^2- d^2)=2K(D+d), from which the result D-d=1/2K follows immediately.

[241] Again, let the given difference be K, and let the two diameters be D and d. Then since D-d=1/2K, D-1/2K=d. Also, since lr.d-d=K, we have lr.d-d=K=1/2K+D-(D-1/2K)=1/2K+D-d, from which it follows that lr.d=1/2K+D.

[242] Using the same notation as in the last note, since D-d=1/2K, 2D-2d=lr.d-d, or 2D=lr.d+d.

[243] Apollonius states the proposition, which applies both to the hyperbola and ellipse, in the form sq.$A\Gamma$:sq.$(BK+T)$::rect.$N\Gamma$,$M\Xi$:sq.$(M\Xi+MN)$, where T is the latus rectum for BK. There is no enunciation; Apollonius continues to rely on the figures in *Conics* VII.6, 7.

times the square on ΘM. Hence, if it be contrived that as the sum of the axis and its latus rectum is to half the proposed sum, so too is that half-sum to some other [line], say ψ; then the line ψ will be given and the rectangle contained by $M\Xi$ and ψ will be equal to the square on ΘM. With that, $M\Xi$ will be to ΘM just as ΘM is to ψ; and, *dividendo*, $\Xi\Theta$ will be to ΘM just as the difference between ΘM and ψ is to ψ: therefore, the rectangle contained by $\Xi\Theta$ and ψ will be equal to what is contained by ΘM and the difference of ΘM and ψ; and, accordingly, that rectangle is given. But it is applied to the line ψ and exceeding by a square, if the axis of the section turns out to be greater than its latus rectum; it is deficient by a square if the latus rectum of the axis turns out to be greater than the axis; whence (*by Lemmas 3 and 4 of the Scholium*) is clear in either case how the synthesis of the problem [is to be carried out].

But as in the preceding [proposition], in this problem as well the synthesis can be set up somewhat better (*paulo paratior habetur compositio*). For since $M\Xi$ is to ΘM just as ΘM is to ψ: by conversion of the ratio and, *permutando*, $M\Xi$ will be to ΘM just as $\Xi\Theta$ is to the excess by which ΘM is greater than ψ: and again, by conversion of the ratio, $M\Xi$ will be to $\Xi\Theta$ just as $\Xi\Theta$ is to the excess by which $\Xi\Theta$ and ψ together are greater than ΘM, in other words (*scilicet*), the excess by which $M\Xi$ is greater than $\Xi\Theta$ [244]; that is, $M\Xi$ will be to $\Xi\Theta$ just as $\Xi\Theta$ is to the excess by which twice $\Xi\Theta$ and ψ together are greater than $M\Xi$, if the latus rectum turns out to be less than the axis. Where the axis turns out to be less than the latus rectum, [we shall have] ratio by ratio (*pari ratione*), as $M\Xi$ is to $\Xi\Theta$ so too will $Z\Theta$ be to the excess by which the difference between ψ and twice $\Xi\Theta$ is greater than $M\Xi$.[245] Therefore, the rectangle contained by $M\Xi$ and the excess just mentioned will equal the square on $\Xi\Theta$. But $\Xi\Theta$ is given. Therefore, that rectangle is given, [and] is applied to a given line, namely, ψ extended or shortened by $\Xi\Theta$, and deficient by a square; whence (*by Lemma 2 of the Scholium*) the line $M\Xi$ will be given, and the point M will be given.

The synthesis will, therefore, be accomplished in the following way. Let it be contrived that as twice the sum of the axis and its latus rectum, or two times $\Gamma\Delta$, is to the given half-sum of the diameter and its latus rectum, so too is the same half-sum to a third proportional, which, for that reason,[246] will equal half the line which we have designated ψ; and let ΘE (placed turned towards A) be made equal to half that same line ψ: whence, ΞE will equal half that [line] to which the rectangle equal to the square on $\Xi\Theta$ is to be applied, and it is the same in either case.[247]

Therefore, let a perpendicular EX be set up equal to $\Xi\Theta$ (*according to Lemma 2*), and with center X and radius ΞE let an arc of a circle be described cutting the axis

---

[244] It is $M\Theta$ which is the excess by which $M\Xi$ is greater than $\Xi\Theta$.

[245] Where the axis is less than the latus rectum, $A\Xi$ is less than $\Xi\Gamma$, by the definition of the homologue. In this case, then, Ξ lies between M and Θ, so that $M\Theta = M\Xi + \Xi\Theta$ instead of $M\Xi - \Xi\Theta$. Therefore, we have the sequence of ratios: $M\Xi{:}\Theta M{::}\Theta M{:}\psi$; by conversion and *permutando*, $M\Xi{:}\Theta M{::}\Theta\Xi{:}\psi - \Theta M$; again, by conversion, $M\Xi{:}\Xi\Theta{::}\Theta\Xi{:}\psi - \Theta M - \Theta\Xi$; finally,

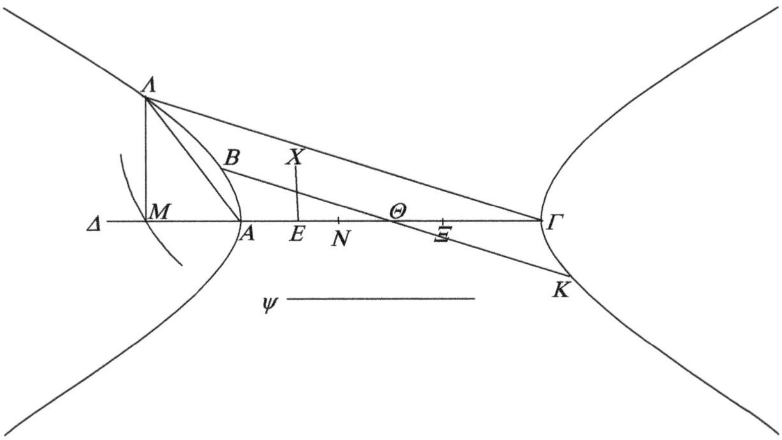

at point M, or at points M, μ, if the problem may be constructed[248] in two ways; as will be shown immediately.

But the limits of this problem will be found (*determinatur autem problema hoc*) from the 38th, 39th, and 40th propositions of [Book] Seven.[249] For if the axis turns out to be greater than its latus rectum, [then] by the 38th [proposition] the sum of the axis and its latus rectum less than any other diameter taken together with its latus

substituting $\Theta\Xi + M\Xi$ in place of ΘM and rearranging the terms slightly, we have $M\Xi{:}\Xi\Theta{::}\Theta\Xi{:}(\psi - 2\Theta\Xi) - M\Xi$.

[246] Let K be the given sum. Then, as argued at the outset, $\Gamma\Delta : 1/2K :: 1/2K : \psi$. Therefore, if $2\Gamma\Delta : 1/2K :: 1/2K$:(third proportional), the third proportional must equal $1/2\psi$.

[247] In the first case, where the axis is greater than the latus rectum, the line to which the square on $\Xi\Theta$ is to be applied is $2\Xi\Theta + \psi = 2\Xi\Theta + 2\Theta E = 2E\Xi$, while in the second case, where the latus rectum is greater than the axis (and, accordingly, where Ξ lies between E and Θ) the line is $\psi - 2\Theta\Xi = 2\Theta E - 2\Theta\Xi = 2E\Xi$.

[248] That is, solved.

[249] *Conics* VII.38 states: "If there is a hyperbola, and the transverse side of the figure constructed on its axis is not less than a third of its *latus rectum*, then the sum of the lines bounding each of the figures on its diameters [i.e., D+lr.D (MNF)] which are not axes is greater than the sum of the lines bounding the figure constructed on its axis; and the sum of the lines bounding the figures constructed on those [diameters] closer to the axis is less than [the sum of] the sides bounding the figures constructed on those farther from it."

Apollonius divides the proof of VII.38 into two parts: VII.38 itself considers the case in which Ax$\geq$lr.Ax while VII.39 takes up the case in which 1/3lr.Ax$\leq$Ax$<$lr.Ax. *Conics* VII.40 goes on to show what happens when Ax$<$1/3lr.Ax.

Specifically, *Conics* VII.40 states: "If there is a hyperbola, and its transverse axis is less than a third of its *latus rectum*, then there are two diameters, [one] on either side of its axis, each of which is equal to one third of its [the diameter's] *latus rectum*, and the sum of the sides bounding the figure constructed on each of the sides bounding the figure constructed on each of the two is less than [the sum of] the sides bounding any one of the figures constructed on the diameters on that side [of the axis]; and the sum of the sides bounding the figures constructed on diameters closer to [that diameter] is less than [the sum of] the sides bounding the figures constructed on [diameters] farther from it."

rectum; it is necessary, therefore, that the sum proposed is greater than the axis and its latus rectum together. Neither is it otherwise if the axis turns out to be less than its latus rectum, but not less than a third of it; for, by the 39th [proposition] of [Book] Seven, it remains the case (*constat*) that the sum of the axis and its latus rectum is less than the sum of any other diameter and its latus rectum; therefore, the proposed sum ought to be greater than [the sum of the axis and its latus rectum]; otherwise the problem is impossible.

But if the axis of the hyperbola turns out to be less than a third of its latus rectum, $\Xi\Theta$ will be greater than a fourth of the axis [250]; and if ΞE be made equal to $\Xi\Theta$,[251] the point E on the axis will fall beyond the vertex A; and with perpendicular EX set up equal to $\Xi\Theta$, a circle with center X, radius ΞE, that is $\Xi\Theta$, will touch the axis at point E; and, accordingly, the diameter BK, whose position in this case is determined (*determinatur*) by E coinciding with M, has an absolute *minimum* (*minimum omnium*) for the sum of it and its latus rectum.

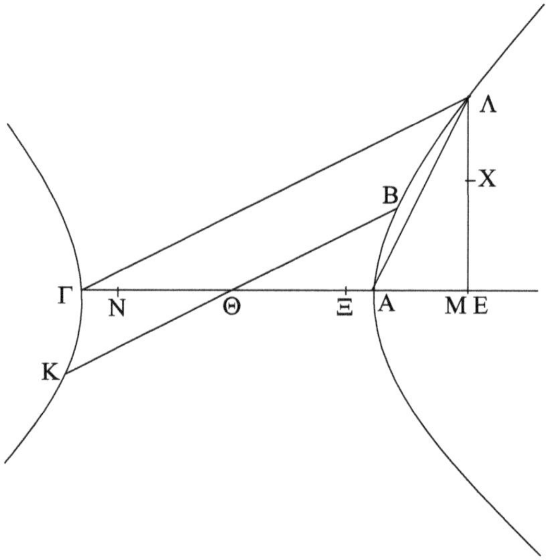

And since NE is three times ΞE,[252] the latus rectum of diameter BK (by VII.6) will be three times BK itself, which is indeed what is fully found in the demonstration of VII.40. Moreover, this diameter BK (by what we showed in the 6th [proposition] of this [work])[253] is the mean proportional between ΞE (or $\Xi\Theta$) and $\Gamma\Delta$, the sum of the axis and its latus rectum; whence the rectangle contained by $\Theta\Xi$, $\Gamma\Delta$

---

[250] $A\Xi<1/3\Xi\Gamma$ or $A\Xi<1/4A\Gamma$. But $A\Theta=1/2A\Gamma$. Therefore, $\Xi\Theta=A\Theta-A\Xi>1/2A\Gamma-1/4A\Gamma=1/4A\Gamma$.

[251] It is precisely this that cannot be done where the axis is greater than a third of its latus rectum. But here one has this freedom, and Halley is taking advantage of this to choose ΞE so that a minimum is obtained.

[252] Since $N\Theta = \Theta\Xi = \Xi E$.

[253] Halley obviously means the 5th proposition.

will be equal to the square on BK. But the sum of the axis and its latus rectum is to the difference of the same just as $A\Theta$ is to $\Theta\Xi$[254]; for which reason, the rectangle contained by the semi-axis $A\Theta$ and the difference of the axis and its latus rectum is equal to the square on BK. And four times the diameter BK is plainly equal to the minimum sum of the diameter and its latus rectum: therefore that minimum sum will be the mean proportion between eight times the axis $A\Gamma$ and the difference between the axis and its latus rectum; and the square of this minimum sum will be equal to eight times the excess by which the figure on the axis is greater than the square of the axis. Accordingly, if eight times this excess is taken away from the square of the sum of the axis and its latus rectum, there will remain the square of the excess by which the latus rectum is greater than three times the axis.[255]

Therefore, for a hyperbola in which the latus rectum of the axis is greater than three times the axis, if it be proposed that a diameter be found which together with its latus rectum produces (*efficiat*) a given sum, and the square of the given sum turns out to be less than the square of the sum of the axis and its latus rectum by an interval (*spatio*) greater than the square of that by which the latus rectum is greater than three times the axis, the problem will be impossible. If, however, the square of the proposed sum together with the square of the excess by which the latus rectum of the axis is greater than three times the axis turns out to equal the square of the sum of the axis and its latus rectum, only the line BK on the same side of the axis[256] fulfills the condition. Here, if the sum proposed turns out to be greater, but less than the sum of the axis and its latus rectum,[257] then the circle described with radius ΞE and center X will cut the axis in two points, M and μ, beyond the vertex A; whence two diameters are obtained, on either side of BK and on the same side of the axis, which have the same sum of themselves and their latera recta: so that four diameters altogether satisfy the condition. If, however, that sum turns out to be equal to the sum of the axis and its latus rectum, [then] besides the axis, two diameters at most (one on either side of [the axis]) will satisfy the problem, the point μ coinciding with the vertex A. But if the proposed sum happens to be greater than that sum, one

---

[254] By definition $A\Xi : \Xi\Gamma$ ::Ax:lr. Therefore, $\Xi\Gamma + A\Xi{:}\Xi\Gamma$::lr+Ax:lr and $\Xi\Gamma + A\Xi$ ($= A\Gamma$):$2\Xi\Gamma$::lr+Ax:2lr, so that $A\Gamma{:}2\Xi\Gamma - (\Xi\Gamma + A\Xi)$ :: $A\Gamma : \Xi\Gamma - A\Xi$ :: $A\Gamma : N\Xi$::lr+Ax: lr−Ax. But $A\Gamma{:}N\Xi$ :: $A\Theta{:}\Theta\Xi$. Therefore, $A\Theta{:}\Theta\Xi$::lr+Ax:lr−Ax.

[255] Since sq.BK = rect.$A\Theta$, lr − Ax we have 16 × sq.BK = 16 × rect.$A\Theta$, lr − Ax, or, since Ax = $2A\Theta$, sq.$4BK$ = rect.8Ax, (lr − Ax). But the minimum sum, $BK + \text{lr}_{BK}$ has been shown to be precisely $4BK$. Therefore, sq.(min. sum) = rect.8Ax, (lr − Ax) = 8 × (rect.Ax, lr − sq.Ax) = 8 × (figure on Ax) − sq.Ax. Therefore, sq.(Ax + lr) − 8 × (figure on Ax) − sq.Ax = sq.(lr + Ax) − 8 × (rect.Ax, lr − sq.Ax = sq.lr − 6 × rect.lr, Ax + 9sq.Ax = sq.lr − 2 × rect.lr, 3Ax + sq.(3Ax) = sq.(lr − 3Ax). Hence, sq.(lr + Ax) − sq.(lr − 3Ax) = sq.(min. sum).

[256] There will naturally be another diameter on the other side, that is, symmetrically positioned with respect to the axis, which together with its latus rectum will equal the given sum; but there will not be another diameter on the same side of the axis satisfying that condition.

[257] Halley's formulation is awkward at best. He must mean that, on the one hand, the previous sum (not the given sum), namely, the square of the excess by which the latus rectum of the axis is greater than three times the axis together with the square of the given sum, is greater than the square of the sum of the axis and its latus rectum, while, on the other hand, the given sum itself is less than the sum of the axis and its latus rectum. For only in this case, according to VII.40, will it be that there are four solutions to the problem, two diameters on either side of the axis.

diameter at most on either side of the axis, beyond BK, will provide a solution; the point μ falling short of vertex A. No *maximum* is given.

Corollary 1. Hence, it is easy to establish that just as diameter BK is one-fourth the sum of BK and its latus rectum,[258] the sum of any two diameters, producing the same sum together with their latera recta, is one-half that common sum.[259]

Corollary 2. Hence, that diameter, which has the same sum with its latus rectum as has some other given diameter, will be equal to half the excess by which the latus rectum of the given diameter is greater than that same diameter; and the latus rectum of that other diameter will equal half the sum of the latus of the given diameter and three times that diameter.[260]

Corollary 3. Moreover, that diameter which has the same sum with its latus rectum as the axis has [with its latus rectum] will be equal to half the excess by which the latus rectum of the axis is greater than the axis. And the latus rectum [for that diameter] will be equal to the half-sum of the latus rectum of the axis and three times the axis. Accordingly, it is less than the latus rectum of the axis by half the excess by which the latus rectum of the axis is greater than three times the axis.

Proposition XXVI (Problem)

Given the axis and latus rectum of an ellipse, it is required to find a diameter which together with its latus rectum makes a given sum.

With the same things we have supposed [before] for the ellipse being kept [the same], consider [the problem] done; and let BK be the diameter which we seek. And so (by the 17th [proposition] of the Seventh [Book]) the square of the axis will be to the rectangle contained by $N\Gamma$, $M\Xi$ just as the square of the given sum of the diameter and its latus rectum is to the square of the sum of MN, $M\Xi$, that is, to the square on $N\Xi$; therefore, by what has been stated many times before, the difference of the axis and its latus rectum will be to $M\Xi$ just as the square of the proposed sum is to the square on $N\Xi$. But the rest is given; consequently, $M\Xi$ is also given. On the other hand, the square of the proposed sum is to the square on $N\Xi$ just as the difference of the axis and its latus rectum is to $M\Xi$: and the point Ξ is given, [so] the point M is also given.

Moreover, the synthesis is clear. For let it be contrived that as the square on the proposed sum is to the square on $N\Xi$, so too is the difference of the axis and its latus rectum to a line equal to $M\Xi$, which is positioned from Ξ towards N; and, if the problem proposed be possible, the point M will fall within the axis AG. But the point M having been obtained, the rest will be carried out as presented before (*ut praemissis*).

---

258 Or $BK + BK$ is one-half the sum of it and its latus rectum.

259 The argument is similar to that justifying corollary 1 of VIII.25; only one uses VII.29 here, where one used VII.30 there.

260 Taking D and d to be the diameters, and d+lr.d, we have, by the first corollary, D+d=1/2(D+lr.D) or d=1/2(lr.D-D). As for the second part, we are given d+lr.d=D+lr.D or lr.d=D+lr.D-d. But, by what was just shown, d=1/2(lr.D-D). Therefore, lr.d=D+lr.D-d=D+lr.D-1/2(lr.D-D)=1/2(lr.D+3D).

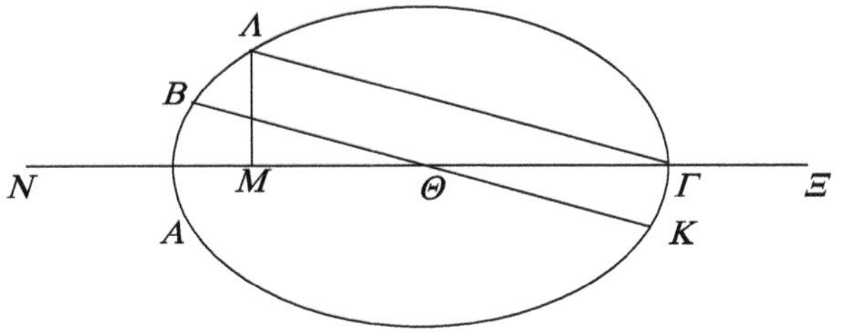

For the limits of this problem, moreover, one appeals to the 41st [proposition] in the Seventh [Book][261]; for that sum can not be less than the major axis and its latus rectum together: nor greater than the sum of the minor axis and its latus rectum. In the former case, the point M will fall beyond vertex A, while in the latter it will fall short of point Γ, outside the ellipse.

But it is possible (*licet*) to find this diameter, given the sum of it and its latus rectum, in an expeditious manner (*expedite*). For by VII.30, the rectangle contained by any chosen (*qualibet*) diameter and the sum of it and the latus rectum is equal to the rectangle contained by the axis and the axis taken together with its latus rectum. Accordingly, the analogy (ἀνάλογον) will be such that as the proposed sum of any diameter and its latus rectum is to the sum of the axis and the latus rectum of the axis, so too will the axis of the ellipse be to the diameter sought; whence it is clear that, because the axis and its latus rectum are given, the sum of any diameter and its latus rectum is reciprocally proportional to the same diameter of the ellipse.[262]

Proposition XXVII (Problem)

Given the sides of the figure of the axis of a hyperbola, it is required to find the position of a diameter which has its figure, or rectangle contained by the diameter and its latus rectum, equal to a given rectangle.

With the same things kept in the figure of the hyperbola presented before, the square on $A\Gamma$ will be (by VII.18) to the rectangle contained by the diameter BK and its latus rectum,[263] just as $N\Gamma$ is to MN. But the square on $A\Gamma$ has been shown to be equal to the rectangle contained by $N\Gamma$ and the sum of the axis and its latus rectum: whence, because $N\Gamma$ is found on both sides,[264] the rectangle contained by MN and

261 *Conics* VII.41 states: "[In] every ellipse, the sum of the [four] sides bounding the figure constructed on its major axis is less than [the sum of] the sides bounding any figure constructed on another of its diameters; and the sum of the sides bounding [one of] the figures constructed on those [diameters] closer to the major axis is less than [the sum of] the sides bounding a figure constructed on [a diameter] farther [from it]; and the sum of the sides bounding the figure constructed on the minor axis is greater than [the sum of] the sides bounding figures constructed on other diameters."

262 Suppose D is the axis sought. Then by VII.30, $\text{rect.}D,(D+\text{lr.}D) = \text{rect.Ax},(\text{Ax}+\text{lr.Ax})$. Therefore, $D : \text{Ax} :: (\text{Ax}+\text{lr.Ax}) : (D+\text{lr.}D)$. By concluding that D and $D+\text{lr.}D$ are reciprocally proportional, that is, referring to the terms D and $D+\text{lr.}D$ only, one wonders whether Halley is actually thinking (here and elsewhere, VIII.23, for example) of a relationship in the form $D = k/(D+\text{lr.}D)$, a quasi-*functional* relationship.

263 As in every analysis, we are supposing the problem done.

264 See note to proposition IX.

the sum of the axis and its latus rectum will be equal to the rectangle or figure proposed. If therefore, that given rectangle is applied to the given line, namely, to $\Gamma\Delta$, the sum of the axis and its latus rectum, the width supplied by the application will be equal to MN, the [line] sought, which is accordingly given: and because of the given point N, the point M is also given.

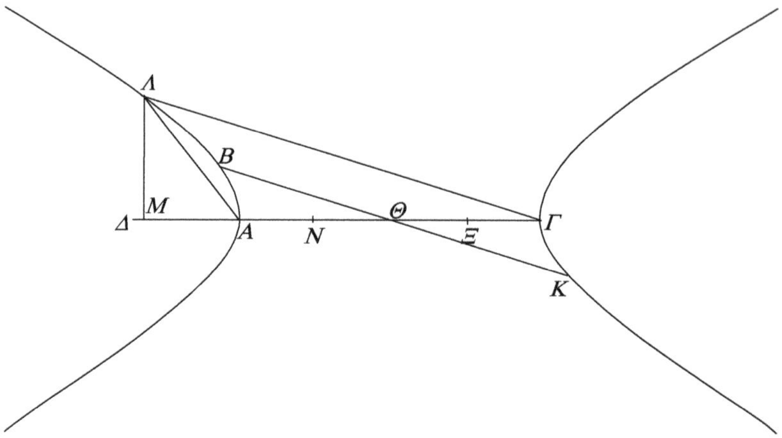

Let the figure proposed be applied, therefore, to the sum of the axis and its latus rectum, and let the width found be placed from point N towards A, as NM. If NM turns out to be greater than NA, the problem will be possible: having found the point M, moreover, the rest will follow as before. But it has a *diorism* from VII.42,[265] from which it is demonstrated that the figure proposed cannot be less than the figure of the axis. The hyperbola has no *maximum* figure.

Proposition XXVIII (Problem)

Given the sides of the figure of the axis of an ellipse, it is proposed to find that diameter of the section which contains with its latus rectum a given figure or rectangle.

With the same things set out in the diagram of the ellipse as before, it may be shown (*probabitur*) (by VII.18), with an argument altogether similar [to the argument in the previous proposition] that the rectangle contained by MN and the difference of the axis and its latus rectum is equal to the figure proposed, or to the rectangle contained by the diameter sought and its latus rectum. So, if that given rectangle be applied to the given line, that is, to the difference of the axis and its latus rectum, the width thus supplied will equal the sought MN, which is, for this reason, given: and the point N being given, the point M is also given.

[265] *Conics* VII.42 states: "The smallest of the figures constructed on diameters of a hyperbola is the one constructed on its axis; and those constructed on [diameters] closer to the axis are smaller than those constructed on [diameters] farther [from it]."

Having applied the rectangle proposed, therefore, to the line $\Gamma\delta$ equal to the difference of the axis and its latus rectum, let MN, placed from N towards the center, be made equal to the width found from the application; and if the point M falls on the axis, or between A and Γ, the problem will be possible: and with the point M given, let perpendicular $M\Lambda$ be set up, whose square is to the rectangle $AM\Gamma$ as the latus rectum of the axis is to the axis[266]; with $\Gamma\Lambda$ joined let BK be drawn, which by what was shown before will be the diameter we are seeking.

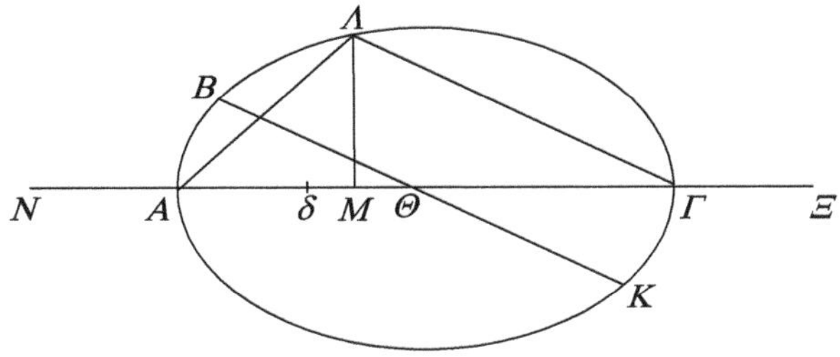

The problem has limits (*limites*), moreover, from the 43rd [proposition] of the Seventh [Book],[267] by which it is established that the figure proposed is not less than the figure of the major axis; otherwise the point M would have fallen short of A, outside the section: nor can it be greater than the figure of the minor axis; for if that turned out to be the case, M would have fallen outside the section beyond Γ. Nor, as we have said many times, is it of any use to find the other diameter equal to BK, at the same distance on the other side of the axis, which would also solve the conditions of the problem.

But in this, as in the preceding [proposition], we can obtain the diameter sought with the help of VII.29 and 30. For since in the ellipse (by the 30th) the sum of the square and the figure of the axis is always equal to the sum of the square of any diameters and its figure, if from the sum of the square of the axis and its figure, the given figure of the diameter sought is taken away, the square of the same diameter will be left. But in the hyperbola the difference of the square and figure of the axis (by VII.29) is equal to the square and figure of any diameter; therefore, the excess by which the square of the axis and the proposed figure together are greater than the figure of the axis will be equal to the square of the diameter sought.

Proposition XXIX (Problem)

Given the sides of the figure of the axis of a hyperbola, it is proposed to find the diameter of the section the square of which together with the square of it latus rectum makes a given sum.

---

[266] This is proven in *Conics* I.21. Why Halley points it out, however, is unclear, for it is hardly necessary.

[267] *Conics* VII.43 states: "The smallest of the figures constructed on diameters of an ellipse is the figure constructed on the major axis; and the greatest of them is the one constructed on the minor axis; and those constructed on [diameters] closer to the major axis are smaller than those constructed on [diameters] farther [from it]."

With the same things described in the diagram of the hyperbola set out as before, consider [the problem] done, and let BK be the diameter we are seeking. Therefore (by VII.19),[268] as the square on $A\Gamma$, or the rectangle contained by $N\Gamma$ and the sum of the axis and its latus rectum, is to the rectangle contained by $N\Gamma$ and $M\Xi$, that is, as the sum of the axis and its latus rectum is to $M\Xi$, so too will the proposed sum of squares on BK and its latus rectum be to the sum of squares on NM and $M\Xi$. And so, applying that given sum of squares to the sum of the axis and its latus rectum, the rectangle contained by $M\Xi$ and the width supplied by the application, let it be ψ, will equal the sum of the squares on MN and $M\Xi$.[269]

Now the axis of the section will be greater than its latus rectum, or less, or equal; so, let it first[270] be greater than it; whence (by VII.6) $M\Xi$ will be greater than MN, and (by *Elem.* II.7) the squares on NM, $M\Xi$ together will be equal to twice the rectangle contained by $NM\Xi$ and the square on $N\Xi$. For this reason, the rectangle contained by $M\Xi$ and the given width ψ just found will equal twice the rectangle contained by $NM\Xi$ and the square on $N\Xi$. But NM is the excess by which $M\Xi$ is greater than $N\Xi$; and, with that, twice the rectangle $NM\Xi$ is equal to twice the excess by which the square on $M\Xi$ is greater than the rectangle contained by $N\Xi M$. Therefore, the square on $N\Xi$ together with twice the excess by which the square on $M\Xi$ is greater than the rectangle $N\Xi M$ is equal to the rectangle contained by ψ and $M\Xi$. And, having removed from both twice that excess, the difference by which the rectangle contained by $M\Xi$ and ψ is greater than twice the excess of the square on $M\Xi$ over the rectangle $N\Xi M$ will equal the square on $N\Xi$: and, halving, the rectangle contained by $M\Xi$ and half ψ and $N\Xi$ together less the square on $M\Xi$ will equal half the square on $N\Xi$.[271] But the square on $N\Xi$ is given: therefore the rectangle contained by $M\Xi$ and $1/2\psi$ and $N\Xi$ together,[272] deficient by the square on $M\Xi$ is given. But [that rectangle] is applied to a given line, namely, the same $1/2\psi$ and $N\Xi$ taken together. Therefore, $M\Xi$ is given, and the point Ξ being given, the point M is also given.

The synthesis, moreover, will be accomplished in this way. With the rest described as before, let a fourth of the proposed sum of squares be applied to $\Gamma\Delta$, the sum of the axis and its latus rectum; and let the width obtained by the application, say ϕ (which is equal to a fourth of ψ), be set along the axis from the center Θ to the [point] E, in the direction of A, so that the rectangle contained by $\Gamma\Delta$, ΘE be

---

[268] *Conics* VII.19 states that sq.$A\Gamma$:(sq.BK+sq.T)::rect.$N\Gamma$,$M\Xi$:(sq.MN+sq.$M\Xi$), where T is the latus rectum for diameter BK. Again, there is no enunciation.

[269] Applying the sum of squares to the sum of the axis and its latus rectum means having found ψ such that rect.(Ax+lr.Ax),ψ=sq.BK+sq.lr.BK. But (Ax+lr.Ax):MX::(sq.BK+sq.lr.BK):(sq.MN+sq.$M\Xi$). Therefore, taking ψ as a common height, rect.(Ax+lr.Ax),ψ:rect.MX,ψ::(sq.BK+sq.lr.BK):(sq.MN+sq.$M\Xi$). And since rect.(Ax+lr.Ax),ψ=sq.BK+sq.lr.BK, it follows that rect.$M\Xi$,ψ=(sq.MN+sq.$M\Xi$).

[270] Proposition XXIX covers this case and, implicitly, also the third case; proposition XXX covers the second case.

[271] Since rect, $M\Xi$,ψ-2(sq.$M\Xi$-rect.$M\Xi$,$N\Xi$)=sq.$N\Xi$, we have 1/2rect, $M\Xi$,ψ- sq.$M\Xi$+ rect.$M\Xi$,$N\Xi$ =1/2sq.$N\Xi$, or rect.$M\Xi(1/2\psi+N\Xi)$-sq.$M\Xi$=1/2sq.$N\Xi$, which is just the form we need to apply the "application of areas" constructions in the last *Scholium*.

[272] The shift to $1/2\psi$ instead of "half ψ" is in Halley's text.

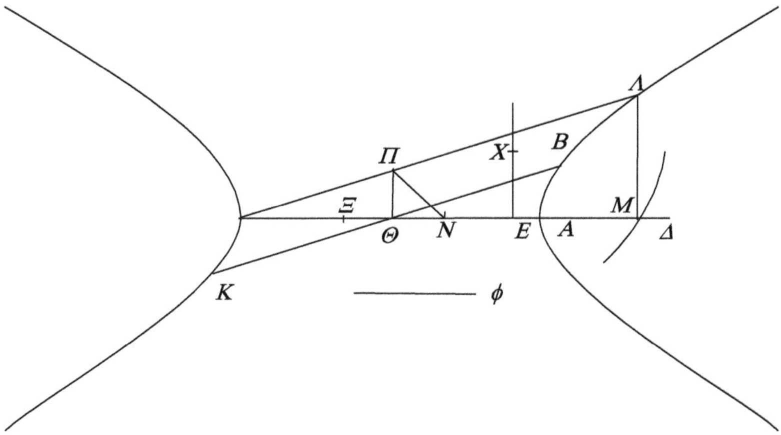

equal to a fourth the given sum of squares; and let *EX* be set up perpendicularly [to the axis]: having made $\Theta\Pi$ along the minor axis equal to $\Theta\Xi$ or ΘN, let $N\Pi$ be joined and let *EX* be taken as equal to the same. Next, with center *X* [and] radius ΞE let a portion of a circle be described, which cuts the axis at point *M*, at which $M\Lambda$ is set up perpendicularly to the axis. The rest is as before. By the reverse of the analysis (*Analysi reciproca*), moreover, the demonstration is sufficiently clear, seeing that *EX* is made equal to that which is equal in square to twice the square on $\Xi\Theta$, or half the square on $N\Xi$. [273]

And there is no other limit than what is demonstrated in VII.41,[274] that the sum of the squares of the axis and of its latus rectum are less than the sum of squares of the sides of the figure of any other diameter. Here, therefore, if the sum proposed turns out to be less, the point *M* will fall short of the vertex, or outside the section; or the circle will not touch the axis; and, accordingly, the problem will be impossible.

Proposition XXX (Problem)

With the same things set out, let the axis of the hyperbola be less than its latus rectum; then it is required to find the diameter of the section for which the squares of the sides of its figure taken together are equal to a given rectangle.

By the same things we demonstrated in the preceding [proposition], from VII.19 as the sum of the axis and its latus rectum is to $M\Xi$, so too will the proposed sum of

---

[273] Indeed, $EX = N\Pi$, while sq.$N\Pi$=sq.$\Theta\Pi$+sq.ΘN=2sq.ΘN=1/2sq.$N\Xi$. Furthermore, the radius of the circle $\Xi E = \Xi\Theta + \Theta E$=1/2($N\Xi$+2$\phi$)=1/2($N\Xi+1/2\psi$). Hence, by the construction given in the 2nd lemma of the *Scholium*, 1/2sq.$N\Xi$=rect.$M\Xi$,($2\Xi E - M\Xi$)=rect.$M\Xi$,($N\Xi+1/2\psi - M\Xi$), that is, 1/2sq.$N\Xi$ has been applied to the line $N\Xi+1/2\psi$, deficient by a square.

[274] *Conics* VII.44 states: "If there is a hyperbola, and the transverse side of the figure constructed on its axis is either [1] not less than its *latus rectum*, or [2] less than it, but [such that] its square is not less than half of the square of the difference between [the transverse side] and [the *latus rectum*]: then the sum of the squares of the two sides of the figure constructed on the axis is less than [the sum of] the squares of the two sides of any figure constructed on one of its other diameters." Hence, the *diorism* is valid both for the case where the axis is greater than its latus rectum—the case treated explicitly here—and also the case where the axis is equal to its latus rectum.

squares be to the sum of squares on *NM* and $M\Xi$: whence, applying this given sum to the sum of the axis and its latus rectum, the rectangle contained by $M\Xi$ and the given width supplied from the application, say line ψ, will be equal to the sum of the squares on *MN*, $M\Xi$.[275] Since the axis $A\Gamma$ is indeed less than its latus rectum, $M\Xi$ will be less than *MN*. Indeed, the squares on *MN* and $M\Xi$ together will (by *Elem.* II.7) be equal to twice the rectangle contained by $NM\Xi$ together with the square on $N\Xi$, that is, twice the rectangle contained by $M\Xi$ and $M\Xi$, $N\Xi$ together, together with the square on $N\Xi$ [276]: wherefore, the rectangle contained by $M\Xi$ and ψ will be equal to twice the rectangle contained by $M\Xi$ and $N\Xi$, $M\Xi$ together, together with square on $N\Xi$: and removing a common [area] from both sides, namely, twice the rectangle contained by $M\Xi$ and $N\Xi$, $M\Xi$ together, the excess by which the rectangle contained by ψ and $M\Xi$ will be greater than twice the rectangle $N\Xi M$, and twice the square on $M\Xi$ together,[277] will equal the square on $N\Xi$: then, taking half of the equality, the rectangle contained by $M\Xi$ and the excess by which ½ψ is greater than $N\Xi$ less the square on $M\Xi$ will be equal to half the square on $N\Xi$.[278] But the square on $N\Xi$ is given; and, with that, the rectangle contained by the aforementioned excess and $M\Xi$ is given. But the rectangle is applied to that given line, namely, the difference of ½ψ and $N\Xi$, and deficient by a square: therefore $M\Xi$ is given, and because the point Ξ is given, *M* is also given.

The synthesis in this case, moreover, differs not at all from the preceding, except that the point Ξ is closer than the center Θ to the vertex *A*. Therefore, having applied a fourth of the proposed sum of squares to the sum of the axis and its latus rectum, let the width thus supplied, say Φ, be placed from Θ towards vertex *A*, as ΘE; and from the point *E* let line *EX* be set up at a right angle to the axis and equal to $N\pi$, which is equal in square to half the square on $N\Xi$ [279]; and with center *X* and radius ΞE let the arc of a circle be described, which indeed, if the problem is possible, will cut the axis beyond the vertex at a point *M*, or at points *M*, μ, under certain conditions to be stated presently.

The problem has a *diorism* from the 45th and 46th [propositions] of the Seventh [Book].[280] For, by the 45th [proposition], if the square of the axis of the hyperbola

---

[275] Same reasoning as in the last proposition.

[276] Since $MN > M\Xi$, we may consider the magnitude $MN - M\Xi$, which is $N\Xi$. Thus the word "indeed" (*verum*) at the start of the sentence. From here, sq.$(MN - M\Xi)$ = sq.MN + sq.$M\Xi$ − 2rect.$MN, M\Xi$ = sq.$N\Xi$ or sq.MN+sq.$M\Xi$ = 2rect.$M\Xi, MN$+sq.$N\Xi$ = 2rect.$M\Xi, (M\Xi + N\Xi)$+ sq.$N\Xi$.

[277] Since 2rect.$M\Xi$,$(M\Xi + N\Xi)$=2sq.$M\Xi$+2rect.$N\Xi, \Xi M$.

[278] Having established that rect.$M\Xi, \psi$ = sq.MN+sq.$M\Xi$, we have rect.$M\Xi, \psi$−2rect.$M\Xi$, $N\Xi$−2sq.$M\Xi$ = sq.$N\Xi$, or ½ rect.$M\Xi, \psi$− rect.$M\Xi, N\Xi$− sq.$M\Xi$ = ½ sq.$N\Xi$, or again, rect.$M\Xi$, $(½\, \psi - N\Xi)$−sq.$M\Xi$ = ½ sq.$N\Xi$.

[279] It is understood that as in the previous proposition π has been marked along the minor axis so that $\Theta\pi = \Theta\Xi = N\Theta$. Therefore, sq.$N\pi$ = 2sq.$N\Theta$ = ½sq.$N\Xi$.

[280] *Conics* VII.45 is, in fact, the second case described in VII.44. *Conics* VII.46, continuing the exposition begun in VII.44, states: "But if the square on transverse diameter [i.e., axis] is less than half of the square of the difference between [the transverse axis] and the *latus rectum* of the figure constructed on it, then on either side of the axis are two diameters, the square on each of which is equal to half of the square on the difference between [the diameter] and the *latus rectum* of the

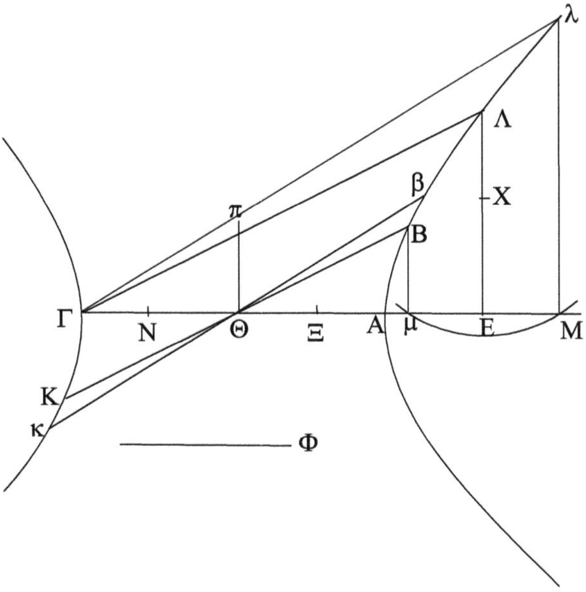

turns out to be not less than half the square of the difference between the axis and its latus rectum, the sum of the squares of the axis and its latus rectum will be less than the squares taken together of the sides of the figure of any other diameter of the section: accordingly, it is required that the proposed sum be greater than the squares of the sides of the figure of the axis; and the greater that sum turns out to be, the farther the diameter, which we are seeking, will diverge (*aberit*) from the axis.

If, in fact, the square of the axis turns out to be less than half the square of the difference between the axis and its latus rectum, it is demonstrated in VII.46 that a diameter will be found on either side of the axis, whose square together with the square of its latus rectum, will be less than the sum of the squares of the sides of the figure of any other diameter taken on the same side of the axis. And diameters closer to that [diameter] on either side have a smaller sum of squares of the sides of the figure than those farther away: for, in this case,[281] the point *E* will fall beyond the vertex of the extended axis of the hyperbola; and *EX*, or that equal in square to half the square on $N\Xi$, will be equal to ΞE, and the circle described with center *X*

figure constructed on it; and the sum of the squares of the two sides of the figure constructed on it is less than [the sum of] the squares of the two sides of any figure constructed on [one of] the diameters drawn on the side [of the axis] on which it lies; and [the sum of] the squares of the two sides of those [figures] constructed on the [diameters] on its side [of the axis] closer to it is less than [the sum of] the squares of the two sides [of figures] constructed on those [diameters] [farther from it]."

[281] Halley is referring to the case of that particular diameter D, for which sq.D+sq.lr.D is a minimum.

will touch the axis at point E, coinciding in this case with the point M: and, with that, a fourth of ψ, or ΘE, will be equal to that which, in square, is twice the square on $\Xi\Theta$ together with $\Xi\Theta$ itself.

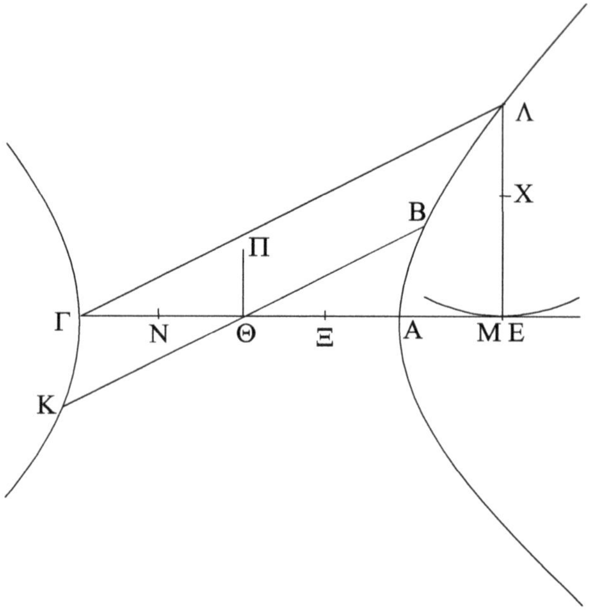

Accordingly, ΘE will be to ΞE just as the diagonal of a square and its side together is to its side, or as $\sqrt{2}+1$ to 1.[282] But the rectangle contained by ΘE, that is, $\frac{1}{4}\psi$, and the sum of the axis and its latus rectum is equal (by construction) to a fourth of the minimum sum of squares; and, accordingly, the rectangle contained by $(\sqrt{2}+1)\times N\Xi$ and the sum of the axis and its latus rectum will be equal to half that minimum sum.[283] But since the sum of the axis and its latus rectum is to the difference of the same just as the axis is to $N\Xi$, the rectangle contained by $N\Xi$ and the sum of the axis and its latus rectum will be equal to the excess by which the figure of the axis is greater than the square of the axis; and, with that, the minimum

[282] Recall ½sq.$N\Xi$ = rect.$M\Xi$,(½ψ − $N\Xi$)−sq.$M\Xi$ or ½sq.$N\Xi$ = rect.$M\Xi$,(2ΘE − $N\Xi$)−sq. $M\Xi$ = rect.$M\Xi$,(2ΘE − $N\Xi$ − $M\Xi$). Since we are assuming a *minimum* solution, we must assume that $M\Xi = 2\Theta E - N\Xi - M\Xi = \sqrt{½} \times N\Xi = \sqrt{2}\times\Xi\Theta$ (which is what Halley is referring to by "that which, in square, is twice the square on $\Xi\Theta$"). Hence, $2\Theta E = N\Xi + M\Xi + \sqrt{2}\times\Xi\Theta = 2\Xi\Theta + 2\sqrt{2}\times\Xi\Theta$ or $\Theta E = \Xi\Theta + \sqrt{2}\times\Xi\Theta = (\sqrt{2}+1)\Xi\Theta$, that is, ΘE:$\Xi\Theta$::$(\sqrt{2}+1)$:1. This, incidentally, also gives further justification for the fact that E, as Halley pointed out, falls "beyond the vertex of the extended axis."

On another front, it should be pointed out that the root signs appear just as they do above in Halley's text.

[283] The line ψ is defined so that rect.ψ,(Ax+lr) = (given sum of squares); however, ¼$\psi = \Theta E = (\sqrt{2}+1)\Theta\Xi = (\sqrt{2}=1)N\Xi/2$. Therefore, ½ (minimum sum of squares) = rect.½ ψ,(Ax + lr) = rect.$(\sqrt{2}+1)N\Xi$,(Ax + lr).

sum of the squares of the sides of the figure will be to the excess by which the figure of the axis is greater than the square of the axis just as $\sqrt{8}+2$ is to unity, or as twice the sum of diagonal of a square and its side is to the side itself.[284]

Therefore, if the proposed sum turns out to be less than this minimum, the problem will be impossible, being that the circle, whose radius is ΞE, will not reach the axis. If it turns out, though, to be greater than [that minimum], but less than the sum of the squares on the sides of the figure of the axis, the circle will cut the axis at two points beyond the vertex, such as at M and μ, by means of which two diameters on either side of the axis may be found in the manner stated many times before, that is, four [diameters] altogether which have the same proposed sum of squares of the sides of the figure. But if the given sum turns out to be equal to the sum of the squares on the sides of the figure of the axis, the point μ will coincide with vertex A; point M, though, will give two other diameters, one on either side of the axis, which have the sum [of the squares on the sides of the figure]. If indeed the proposed sum turns out to be greater than the sum of the squares of the sides of the figure of the axis, the point μ will fall short of vertex A : whereas the other intersection point M will produce two diameters one on either side [of the axis] which satisfy the problem. But, by the nature of the hyperbola, no *maximum* sum of squares can be given.

But if the axis turns out to equal its latus rectum, every diameter will equal its latus rectum (by VII.23)[285]; and, with that, half the proposed sum will equal the square of the diameter we seek.[286]

Corollary 1. Hence, it can be clearly established that the squares of two diameters, having the same sum of squares on the sides of their figures, taken together will equal the excess by which the half-sum of those squares of the sides of the figure, together with the square of the axis, is greater than the figure of the axis.[287]

---

[284] By definition $A\Xi$:$\Xi\Gamma$::Ax:lr. Therefore, $A\Gamma(=\text{Ax})$:$N\Xi$::$(\Xi\Gamma+A\Xi)$:$(\Xi\Gamma-A\Xi)$::(lr + Ax): (lr-Ax) or rect.Ax,(lr − Ax) = rect.Ax,lr − sq.Ax = rect.$N\Xi$,(lr + Ax). But it has just been shown that $2(\sqrt{2}+1)$rect.Ax,lr − sq.Ax = $2(\sqrt{2}+1)$rect.$N\Xi$,(lr + Ax) = min. sum of squares, so, since $2(\sqrt{2}+1)=\sqrt{8}+2$, we have (min. sum of squares):(rect.Ax,lr − sq.Ax)::$(\sqrt{8}+2)$:1.

[285] *Conics* VII.23, stated above in VIII.7, refers to the equality of conjugate diameters; but, as Apollonius himself points out in VII.23, if the conjugate diameters are equal, then (by the definitions following *Conics* I.16) the latera recta must be equal to the diameters as well.

[286] This of course provides only the magnitude of the diameter; its position can then be found via Proposition V.

[287] The pattern of the argument here runs parallel to that in the first corollary to proposition XXIV and to proposition XXV above. But while those arguments were probably algebraic, as I said, the argument here (and in the first corollary to the next proposition) is undoubtedly algebraic. Let the diameters be D and d, let their latera recta be lr.D and lr.d, and the common sum of the squares of the sides of the figure be K, that is, $D^2+(\text{lr.D})^2=d^2+(\text{lr.d})^2=K$. Call the latter the basic condition. Squaring $D^2+(\text{lr.D})^2$ and $d^2+(\text{lr.d})^2$, we obtain $(D^2)^2+2D^2(\text{lr.D})^2+((\text{lr.D})^2)^2=(d^2)^2+2d^2(\text{lr.d})^2+((\text{lr.d})^2)^2$. Rearranging and factoring, we have

$$(D^2-d^2)\,(D^2+d^2)=2[d^2(\text{lr.d})^2-D^2(\text{lr.D})^2]+[(\text{lr.D})^2-(\text{lr.d})^2][(\text{lr.D})^2+(\text{lr.d})^2]\ (*).$$

Now, from the basic condition two other relations follow immediately, namely, $(\text{lr.d})^2-(\text{lr.D})^2=D^2-d^2$ and $(\text{lr.D})^2+(\text{lr.d})^2=2K-(D^2+d^2)$. Substituting these into (*) and canceling the common factor D^2-d^2 (remember, we are considering two diameters on the same side of the axis for which the sum of squares of the sides of the figures are equal), we have

Corollary 2. Accordingly, the square of that diameter, which has its sum [of the squares of the sides of its figure] equal to [the sum] for the axis, will equal the excess by which the half-sum of the squares of the sides of the figure are greater than the figure of the axis, that is, half the square of the difference of the axis and its latus rectum. Moreover, the square of its latus rectum will equal the same half-sum of squares of the figure of the axis taken together with the same rectangle or figure of the axis; that is, half the square on the sum of the axis and its latus rectum.[288]

Corollary 3. The same is to be said for any other given diameter which is not the axis of the section.[289]

Proposition XXXI (Problem)

Given the sides of the figure of the axis of an ellipse, it is required to find a diameter square of which taken together with the square of its latus rectum makes a proposed sum.

With those things described the previous diagrams for the ellipse kept [as before], consider [the problem] done, and let BK be that diameter we are seeking. Therefore, (by VII.19) as the square on $A\Gamma$, or the rectangle contained by $N\Gamma$ and the difference of the axis and its latus recum, is to the rectangle contained by $N\Gamma$ and $M\Xi$, that is, as $\Gamma\delta$, the difference of the axis and its latus rectum, is to $M\Xi$, so too will the proposed sum be to the sum of squares on MN, $M\Xi$; and, with that, having applied a rectangle equal to the proposed sum to the difference of the axis and latus rectum, a width ψ is thence specified, which will, accordingly, be given. It is clear then that the rectangle contained by $M\Xi$ and ψ is equal to the squares on MN, $M\Xi$ together. Indeed, (by *Elem.* II.4) the square on $N\Xi$ is equal to the squares on MN, $M\Xi$ together with twice the rectangle contained by $NM\Xi$, that is, the square on $N\Xi$ is equal to the rectangle contained by $M\Xi$ and ψ together with twice the rectangle contained by $NM\Xi$; but the rectangle contained by $NM\Xi$ is equal to the rectangle contained by $N\Xi M$ less the square on $M\Xi$: wherefore the rectangle contained by $M\Xi$ and both $N\Xi$ and $\frac{1}{2}\psi$ together, less the square on $M\Xi$, is equal to half the square on $N\Xi$. But the square on $N\Xi$ is given. Therefore, the rectangle contained by $M\Xi$ and both $N\Xi$ and $\frac{1}{2}\psi$ together, less the square on $M\Xi$ is given. Moreover, it is applied to a given line, namely, that equal to $\frac{1}{2}\psi$ and $N\Xi$ taken together, deficient by a square. Therefore, the line $M\Xi$ is given, and M is given as well.

$D^2+d^2=2K-(D^2+d^2)-2[D(lr.D)+d\ (lr.d)]$. From here, we add $2(D^2+d^2)$ to both sides and rearrange to obtain $4(D^2+d^2)=2K+2[D^2-D(lr.D)+d^2-d(lr.d)]$ or $D^2+d^2=\frac{1}{2}K+\frac{1}{2}[D^2-D(lr.D)+d^2-d(lr.d)]$. But, by VII.29, $D^2-D(lr.D)=d^2-d(lr.d)=Ax^2-Axlr.Ax$. Therefore, $D^2+d^2=\frac{1}{2}K+\frac{1}{2}[2(Ax^2-Axlr.Ax)]=\frac{1}{2}K+Ax^2-Axlr.Ax$.

288 In this case, $D^2+(lr.D)^2=Ax^2+(lr.Ax)^2=K$. Therefore, $D^2+Ax^2=\frac{1}{2}K+Ax^2\text{-}Axlr.Ax$, or $D^2=\frac{1}{2}K\text{-}Axlr.Ax=\frac{1}{2}[Ax^2+(lr.Ax)^2\text{-}Axlr.Ax]=\frac{1}{2}[Ax^2+(lr.Ax)^2\text{-}2Axlr.Ax]=\frac{1}{2}(Ax\text{-}lr.Ax)^2$. As for the second conclusion, from the basic condition, $D^2+(lr.D)^2=Ax^2+(lr.Ax)^2$, it follows that $(lr.D)^2=Ax^2+(lr.Ax)^2\text{-}D^2$. Substituting $D^2=\frac{1}{2}[Ax^2+(lr.Ax)^2\text{-}2Axlr.Ax]$, we have $(lr.D)^2=Ax^2+(lr.Ax)^2\text{-}\frac{1}{2}[Ax^2+(lr.Ax)^2\text{-}2Axlr.Ax]=\frac{1}{2}[Ax^2+(lr.Ax)^2+2Axlr.Ax]=\frac{1}{2}(Ax+lr.Ax)^2$.

289 Since by VII.29, $\Delta^2\text{-}\Delta lr.\Delta=Ax^2\text{-}Axlr.Ax$ for any diameter Δ, $D^2+d^2=\frac{1}{2}K+Ax^2\text{-}Axlr.Ax=\frac{1}{2}K+\Delta^2\text{-}\Delta lr.\Delta$. In other words, the diameter Δ can replace the axis in the previous two corollaries.

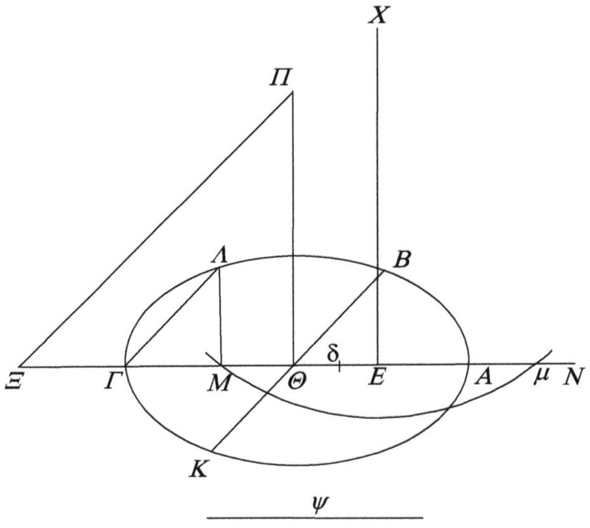

The synthesis of the problem will be accomplished in this manner. Let a fourth of the given sum of squares be applied to the difference of the axis and its latus rectum; and with the width found, or a fourth of ψ, set [line] ΘE equal [to it] along the axis from the center Θ towards N. And from the point E let EX be set up perpendicularly [with respect to the axis] equal to $\Xi\Pi$, or that equal in square to half the square on $N\Xi$ [290]: next, with center X and radius XE a circle is described, which, if the problem be possible, cuts the axis between A and Γ at a point M; and, having set up $M\Lambda$ perpendicular to the axis, both the magnitude and position of the sought diameter, BK, are obtained.

But the limits of this problem will be found from VII.47 and 48.[291] For if the square of the major axis of the ellipse turns out to be not greater than half the square

[290] The construction is as in the previous propositions, namely, a segment $\Theta\Pi$, equal to $\Xi\Theta$, is cut off the minor axis extended, and $\Xi\Pi$ is joined: thus, sq.$\Xi\Pi$=2sq.$\Theta\Xi$=½sq.$N\Xi$.

[291] *Conics* VII.47 and 48 are the analogues for the ellipse of VII.45 and 46. Their statements are as follows:

Conics VII.47: "If there is an ellipse, and the square on the transverse side of the figure constructed on its major axis is not greater than half of the square on the sum of the two sides of the figure constructed on it, then [the sum of] the squares on the two sides of the figure constructed on the major axis is less than [the sum of] the squares on the two sides of [all] other figures constructed on its diameters; and [the sum of] the squares on the two sides of those [figures] constructed on [diameters] closer to it [the major axis] is less than [the sum of] the squares on the two sides [of those figures] constructed on [diameters] farther [from it]; and the greatest of them is [the sum of] the squares on the two sides of the figure constructed on the minor axis."

Conics VII.48: If there is an ellipse, and the square on its major axis is greater than half of the sum of the [square (MNF—the Arabic text has "square*s*," which is wrong of course, and was corrected by Halley in his own edition)] on the two sides of the figure constructed on [the major axis], then there are two diameters, [one] on either side of the axis, such that the square on each of them is equal to half of the square on the sum of the two sides of the figure constructed on it; and [the sum of] the squares on the two sides of the figure constructed on it is less than [the sum of] the squares on the two sides of [any of] the other figures constructed on diameters drawn in that quadrant in which [that diameter] is; and [the sum of] the squares on the two sides of [figures] of the sum of the sides of figure of the axis, it is clear (by what is said in the 47th [proposition]) that the proposed sum of squares cannot be less than the squares taken together of the sides of the figure of the major axis, in which case the point M will fall short of vertex A; nor [can it be] greater than the squares taken together of the sides of the figure of the minor axis: for in that case (*posito*) the point M will fall beyond vertex Γ, and the problem will be impossible. If, however, the square of the major axis turns out to be equal to half the square of the sum of the sides of its figure, and the sum is proposed to be equal to the sum of the squares of the sides of the axis, the point E will coincide with the point A. In the remaining cases E will be removed far from the center Θ and will fall outside the section.

If, however, the square on the axis $A\Gamma$ turns out to be greater than half the square on the sum of the sides of the figure of the axis, [then] (by VII.48) a diameter will be obtained on either side of the axis whose square will be equal to half the square on the same diameter and its latus rectum taken together; indeed, the square of this diameter together with the square on its latus rectum makes a sum which is an absolute *minimum*: and the squares of the sides of the figure of a diameter closer to that [diameter] on either side will be less than the squares of the sides of the figure of a diameter farther from the same, exactly as was demonstrated there [in VII.48]. What that minimum turns out to be will be immediately clear by the same argument we used in [the case of] the hyperbola. For since the point M, in the case of the *minimum sum*, coincides with the point E; XE will be equal to line ΞE, which is always equal to that which is in square twice the square on $N\Theta$; and with that ΘE will be equal to the excess by which that equal in square to twice the square on $N\Theta$ is greater than $N\Theta$[292]: whence, ΘE will be to $N\Theta$ just as the excess by which the diagonal of a square is greater than its side to the side itself, or as $\sqrt{2}-1$ is to 1.[293]

By the construction, moreover, the rectangle contained by ΘE and the difference of the axis and its latus rectum is equal to a fourth of the sum of squares of the sides of the figure; and, with that, the rectangle contained by $(\sqrt{2}-1)\times N\Xi$ and the difference of the axis and its latus rectum will be equal to half the minimum sum of squares of the sides of the figure, which we are seeking.[294] But since the difference of the axis and its latus rectum is to the sum of the same [i.e., the axis and its latus rectum] just as the axis itself is to $N\Xi$, the rectangle contained by $N\Xi$ and the difference of the axis and its latus rectum will be equal to the rectangle contained by the axis and the sum of the axis and its latus rectum, that is, the square of the axis together with its figure, or the sum of the squares on each axis[295]: therefore, the

constructed on those diameters in that quadrant closer to it is less than [the sum of] the squares on the two sides of [figures] constructed on those farther [from it]."

292 Again, "that equal in square to twice the square on $N\Theta$," is the line L such that sq.L=2sq.$N\Theta$ or L=$\sqrt{2}N\Theta$. So, QE=$\sqrt{2}N\Theta$-$N\Theta$=($\sqrt{2}$-1)$N\Theta$.

293 The square root signs appear in Halley's text.

294 For this, and what follows, see the notes for the corresponding section in the previous proposition.

295 Since by *Conics* I.15, Ax: ax::ax:lr.Ax (where ax denotes the minor axis, and Ax, as always, denotes the major axis) the figure on the major axis is equal to the square on the minor axis. Therefore, sq.Ax+rect.Ax,lr.Ax=sq.Ax)+sq.ax.

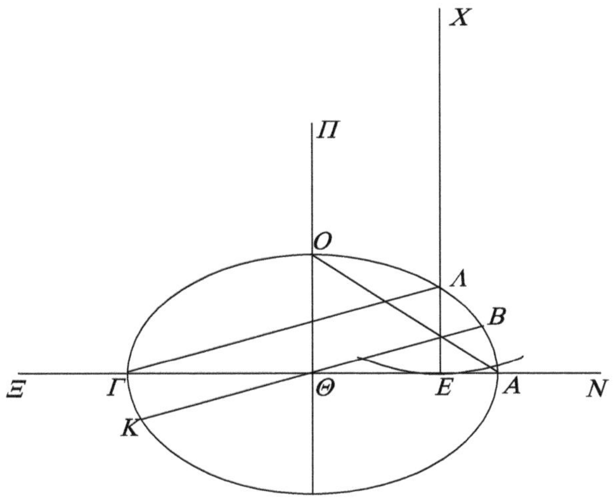

minimum sum of squares on the sides of the figure will be to the sum of the squares on the axes of the ellipse just as $\sqrt{8}-2$ is to unity, or as twice the excess by which the diagonal of a square is greater than its side is to the side itself.

Wherefore, if it is proposed that one search for a diameter in an ellipse which has a sum of squares of the sides of [its] figure less than that just obtained,[296] the problem will be impossible, and the circle just described will not reach the axis. If, however, [the sum] turns out to be greater than the *minimum* stated, but less than the sum of the squares of the sides of the figure of the [major] axis, the circle will meet the axis at two points M, μ within the section. Accordingly, on either side of the axis there will be two diameters, that is, four altogether, which have the proposed sum of squares of the sides of the figure. And if it turns out that the proposed sum is equal to the squares on the axis and on its latus rectum taken together, then the axis itself and two other diameters, one on either side [of the axis], will satisfy the condition.[297] But again, if a sum greater [than the sum with respect to the major axis] is proposed, but it is less than the sum of the squares on the sides of the figure of the minor axis, one diameter on either side of the axis will be found which satisfies the problem, the point M falling within the section. Moreover, the maximum sum of squares on the sides of the figure of the minor axis; which is to the sum of the squares on the sides of the figure of the major axis in the ratio of the [major] axis to its latus rectum.[298]

296 That is, the minimum obtained in the previous paragraph.

297 Namely, that the sum of squares of the sides of their figures are both equal to the sum of the squares of the sides of the figure of the axis.

298 By *Conics* I.15, Ax:ax::ax:lr.Ax and ax:Ax::Ax:lr.ax, from it follows, sq.Ax:sq.ax:: sq.ax: sq.(lr.Ax) and sq.ax:sq.Ax::sq.Ax:sq.(lr.ax). Therefore, sq.Ax:sq.ax::sq.ax:sq.(lr.Ax)::sq.(lr.ax): sq.Ax::[sq.ax+sq.(lr.ax)]:[sq.(lr.Ax)+sq.Ax]. But again, from the proportion Ax:ax::ax:lr.Ax, it follows that sq.Ax:sq.ax::Ax:lr.Ax; therefore, Ax:lr.Ax::[sq.ax+sq.(lr.ax)]:[sq.(lr.Ax)+sq.Ax]. From ax:Ax::Ax:lr.ax, it follows that sq.ax:sq.Ax::ax:lr.ax, so that, in this case, ax:lr.ax:: [sq.Ax+sq.(lr.Ax)]:[sq.(lr.ax)+sq.ax].

If what is proposed is greater than this, the problem will again be impossible, the point M ending up beyond the vertex Γ.

Corollary 1. The sum, moreover, of the squares of any two diameters, having the same sum of squares of the sides of the figure, is equal to half the proposed sum together with the rectangle contained by the axis and the axis and its latus rectum taken together, or, rather, with the squares of the axes taken together.[299]

Corollary 2. And, accordingly, that diameter which has the same sum of squares as the axis itself has,[300] will be equal in square to that which is half the square on the axis and its latus rectum taken together: but its latus rectum equals in square half the square of the excess by which the major axis is greater than its latus rectum. And the same is true even if the given diameter turns out not to be the axis.[301]

Corollary 3. Whence it is clear that those four diameters of the ellipse, to the extent there are four, which can have, as has been said, the same sum, always fall between the major axis and the equal conjugate diameters; since these diameters are greater than their latera recta.[302]

Corollary 4. But the square of that diameter, which has the absolute minimum sum of squares, will be to that which is equal in square to half the sum of the squares of the axes, that is, to the square on AO, as the diagonal of the square to its side, or as $\sqrt{2}$ is to 1: and the latus rectum of the diameter to the diameter itself will be as $\sqrt{2}$ to $2 - \sqrt{2}$.[303]

Corollary 5. Whence in every ellipse, both the diameter and latus rectum, of which the sum of squares is a minimum, will have always the same ratio to AO subtending the quadrant of the ellipse.

Corollary 6. And it is clear, from the limits already stated, that if the major axis has to the conjugate axis the ratio of unity to $\sqrt{(\sqrt{2} - 1)}$, or 1 to 0.6436; it will be possible to draw four diameters, which have the same sum of squares of it and its latus rectum [*sic*] and not otherwise.

---

299 The argument is along the same lines as in the first corollary of the last proposition, only here one uses VII.30, where in proposition XXX one used VII.29.

300 The expressions "same sum of squares" or "same sum" refer to the sum of the squares of the diameter and its latus rectum.

301 Again, the arguments for all three statements are completely analogous to those which establish the corollaries at the end of proposition XXX.

302 In order for there to be four diameters having the same sum of squares of the sides of the figure, the sum must be less than that of the squares of the sides of the figure of the major axis, as stated above and in VII.48. But for a diameter D having this sum equal to the sum of the sides of the figure of the major axis, we have from the previous corollary, sq.D=½sq.(Ax+lr.Ax), while sq.(lr.D)=½sq.(Ax-lr.Ax); therefore, diameter D is still greater than its latus rectum, and, accordingly, must lie closer to the major axis than that diameter which is *equal* to its latus rectum. The latter, however, is precisely the diameter which is equal to its conjugate: for by *Conics* I.15, sq.D:sq.D′::D:lr.D for any diameter D and its conjugate D′; therefore, D=lr.D if and only if sq.D=sq.D′, that is, if and only if D=D′.

303 Corollaries 4, 5, and 6 are not entirely transparent.

Proposition XXXII (Problem)

Given the axis and latus rectum of a hyperbola, find a diameter whose square differs from the square of the latus rectum by a given difference.

As described in the previous [propositions], consider [the problem] done: so let BK be the diameter sought, whose square differs from the square of its latus rectum by a given difference. By VII.20,[304] the square of the axis of the section, or the rectangle contained by $N\Gamma$, $\Gamma\Delta$, will be to rectangle contained by $N\Gamma$, $M\Xi$, that is, $\Delta\Gamma$ to $M\Xi$, just as the proposed difference of squares to the difference of squares on NM, $M\Xi$. But the difference of the squares on NM, $M\Xi$ (by *Elem*.II.6) is equal to four times the rectangle contained by ΘM, $\Theta\Xi$[305]; and, with that, if a rectangle contained by $\Gamma\Delta$ and some other line ψ be made equal to a fourth of the proposed difference of squares, the line ψ will be given: and, by the argument which we have so often adopted (*toties usurpato*), the rectangle contained by $M\Xi$ and ψ will be equal to the rectangle contained by ΘM, $\Theta\Xi$; the analogy (ἀνάλογον), therefore, is that as $\Theta\Xi$ is to ψ, so too will $M\Xi$ be to ΘM; and, *dividendo*, the difference of $\Theta\Xi$ and ψ will be to $\Theta\Xi$ just as $\Theta\Xi$ is to ΞM. But $\Theta\Xi$ and ψ are given, so $M\Xi$ will also be given, whence the point M is given.

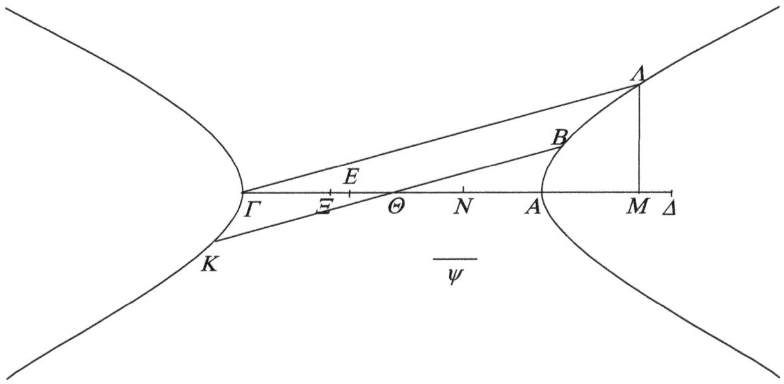

Therefore, the synthesis of the problem will be accomplished if a fourth of the given difference be applied to $\Gamma\Delta$, or to the sum of the axis and its latus rectum; whence the width will be made equal to ψ. Next, let ΘE, equal to ψ, be placed along the axis in the direction of Ξ, and let it be contrived that as $E\Xi$ is to $\Xi\Theta$ so too will $\Theta\Xi$ be to ΞM. But having found point M the diameter BK will also be given, both in magnitude and position; as has been stated so much in the preceding [pages].

---

[304] *Conics* VII.20 applies both to the hyperbola and ellipse and states that sq.$A\Gamma$:(sq.$BK-$ sq.T)::rect.$N\Gamma$,$M\Xi$:(sq.$\Xi M-$sq.MN), where T is the latus rectum for BK (the differences of squares, of course, are reversed where $BK < T$). As in VII.17 mentioned above, there is no enunciation, and Apollonius continues to rely on the figures in *Conics* VII.6, 7

[305] For sq.$\Xi M-$sq.MN=rect.($\Xi M+MN$, $\Xi M-MN$) by *Elem*.II.6, and $\Xi M+MN = MN+N\Xi+MN = 2MN+2N\Theta = 2M\Theta$, while $\Xi M-MN = \Xi N = 2\Theta\Xi$.

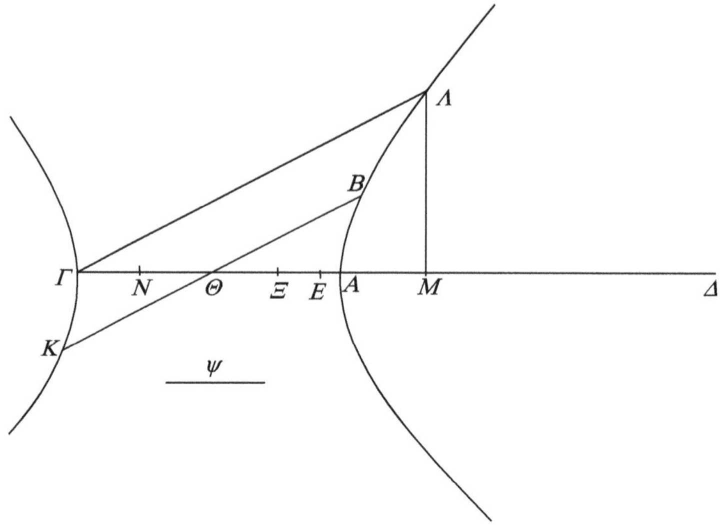

The *diorism* for this problem appeals to VII.49 and 50.[306] For if the axis turns out to be greater than its latus rectum, the difference of the squares of the axis and its latus rectum will be less than any other difference of squares on a diameter and [its] latus rectum: and, with that, the proposed difference cannot be less than that [the difference between the squares of the axis and its latus rectum], as established by the 49th proposition [of Book VII]. But if the axis turns out to be less than its latus rectum, the difference of the squares of the axis and its latus rectum will be greater than any other arbitrary difference [of the squares of a diameter and its latus rectum]; accordingly, the proposed difference cannot be greater than the excess by which the square of the latus rectum of the axis is greater than the square of the axis itself, exactly as set out in VII.50.

[306] *Conics* VII.49 states that, "If there is a hyperbola, and the transverse side of the figure constructed on its axis is greater than its *latus rectum*, then the difference between the squares on the two sides of that figure is less than the difference between the squares on the two sides of any of the figures constructed on the other diameters; and the difference between the squares on the two sides of those [figures] constructed on [diameters] closer [to the axis] is less than the difference between the squares on the two sides of those constructed on [diameters] farther [from it]; and the difference between the squares on the two sides of any of the figures constructed on diameters which are not axes is greater than the difference between the square on the axis and the figure constructed on it, but less than twice that [difference].

Conics VII.50 states that, "If there is a hyperbola, and the transverse side of the figure constructed on its axis is less than its *latus rectum*, then the difference between the squares on the two sides of the figure constructed on the axis is greater than the difference between the squares on the two sides of any of the figures constructed on diameters other than it; and the difference between the squares on the two sides of those [figures] constructed on [diameters] closer to the axis is greater than the difference between the squares on the two sides of those constructed on [diameters] farther [from it]; and the difference between the square on any of those diameters and the square on the *latus rectum* of the figure constructed on it is greater than twice the difference between the square on the axis and the figure constructed on the axis."

Again, by the same [propositions] it is demonstrated, in the former case, that the *maximum* difference of squares is not greater than twice the excess by which the square of the axis is greater than its figure, and, in the latter case, that the *minimum* difference is not less than twice the excess by which the figure of the axis is greater than the square of [the axis]. For, the points E and Ξ merging [in these cases], the line ΞE vanishes, and the line ΞM goes off to infinity (*in infinitum abit*).

Proposition XXXIII (Problem)

Given the axis and latus rectum of an ellipse, find diameters of it whose squares exceed the squares of their latera recta by a given difference, and fall short by the same.

With those things in the ellipse which have been described until now kept the same, the square on $A\Gamma$ or the rectangle contained by $N\Gamma,\Gamma\delta$ will be to the rectangle contained by $N\Gamma,M\Xi$ (by VII.20) just as the given difference of square on BK and its latus rectum[307] is to the difference of squares on NM, $M\Xi$: and, with that, as stated so many times before, $\Gamma\delta$ will be to $M\Xi$ just as the proposed difference is to the difference of squares on NM, $M\Xi$, that is, (by *Elem*.II.5) to four times the rectangle contained by $\Theta\Xi$, ΘM. Accordingly, if a fourth of the given difference of squares be applied to the line $\Gamma\delta$ and it has a width that we call ψ, that is, if it be contrived the rectangle contained by $\Gamma\delta$ and ψ be equal to a fourth of the given difference, [then] the rectangle contained by $M\Xi$ and ψ will be equal to the rectangle contained by $\Theta\Xi$ and $M\Xi$: the analogy (ἀνάλογον) is, therefore, that as ψ is to $\Theta\Xi$ so too will ΘM be to $M\Xi$: but, *componendo* and *dividendo*, $\Theta\Xi$ plus and minus the line ψ will be to $\Theta\Xi$ just as $\Theta\Xi$ is to ΞM.[308] So, $\Theta\Xi$ and ψ are given; whence, because Θ is given, the point M is also given.

Therefore, the synthesis of the problem will be accomplished if a rectangle contained by the axis and its latus rectum and some other [line] ψ be made equal to a fourth the given difference of squares; and, on either side of the center, let lines ΘE, $\Theta\varepsilon$ equal to ψ be placed [along the axis]. Next, let it be contrived that as $E\Xi$ is to $\Xi\Theta$ so too is $\Theta\Xi$ to ΞM, when the diameter is greater than its latus rectum; and [let it be that] as $\varepsilon\Xi$ is $\Xi\Theta$ so $\Theta\Xi$ is to $\Xi\mu$, when the latus rectum turns out to be greater than the diameter: having found the points M, μ, we will find each diameter as before. Nor is there any point to go back and establish the result, it being as clear as can be from the analysis.

From those things related in the final proposition of the Seventh Book,[309] moreover, one can obtain the limits for this problem. For among all the diameters of the

---

[307] Thus, as before, we are assuming the problem has been solved, and BK is the required diameter.

[308] It is *componendo* or *dividendo*, and, accordingly, plus or minus, depending on whether M is between Θ and Ξ or Θ and N.

[309] *Conics* VII.51 states: The difference between the squares on the two sides of the figure constructed on the major axis of an ellipse is greater than the difference between the squares on the two sides of any figure constructed on other diameters which are greater than the *latus rectum* of the figures constructed on them [the diameters]; and the difference between the squares on the two sides of those [figures] constructed on those diameters closer to the major axis is greater than the difference between the squares on the two sides of those constructed on those farther [from it]; and the difference between the squares on the two sides of the figure constructed on its minor axisis

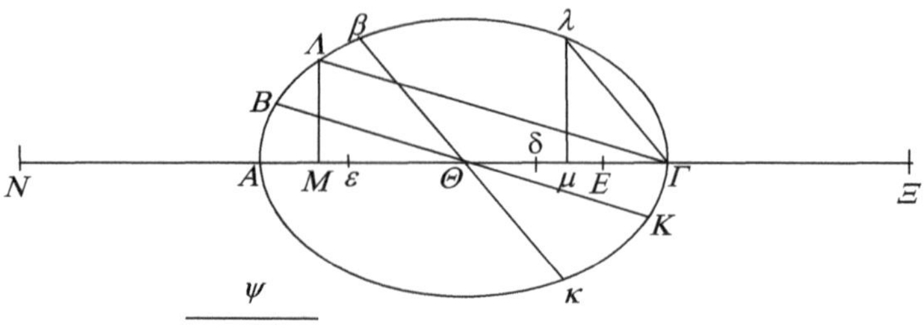

ellipse which are greater than their latera recta, the square of the major axis exceeds the square on its latus rectum by a greater interval; and among those which are less than their latera recta, the difference of those squares [with respect to] the minor axis is the *maximum* of all; indeed, the difference [between the square of the latus rectum of the minor axis and the square of the minor axis] is greater than the excess by which the square of the major axis is greater than the square of its latus rectum in the ratio of the latus rectum of the major axis to the axis itself.[310] For this reason,[311] if the given difference turns out to be less than the difference of the squares of the major axis and its latus rectum, there will be four different diameters, two on either side of the axis, that satisfy the problem, being that each of the points M and μ will fall between the vertices A, Γ, of the ellipse. But if it turns out to be greater than that,[312] but less than the difference of the squares of the minor axis and its latus rectum, then two diameters only will exhibit the [given difference], one on either side of the minor axis. But if it turns out to be greater than that as well, then the problem will be impossible, each of the points M, μ falling outside the axis $A\Gamma$. But there is no minimum for the given difference of squares; for in equal conjugate diameters this difference comes out to zero (*nulla*) since the latera recta become equal to the diameters themselves [in this case].

Since also ΞE is to $\Xi\Theta$ just as $\Xi\Theta$ is to ΞM, $\Xi\Theta$ will be to ΘE just as ΞM is to $M\Theta$: and by the same reasoning $\Xi\Theta$ will be to $\Theta\varepsilon$, that is, to ΘE, just as $\Xi\mu$ is to $\mu\Theta$; and, with that, ΞM will be to $M\Theta$ just as $\Xi\mu$ is to $\mu\Theta$. For which reason, in every case ΞM is divided harmonically at points Θ, μ; and, accordingly, whatever diameter is given, it will be easy to find the corresponding [diameter] which has the same difference of squares of itself and its latus rectum.

---

greater than the difference between the squares on the two sides of any figure constructed on other diameters which are smaller than the *latera recta* of the figures constructed on them; and the difference between the squares on the two sides of the [figures] constructed on those of these diameters closer to the minor axis is greater than the difference between the squares on the two sides of those constructed on [diameters] farther from it."

[310] This follows entirely by *Conics* I.15: Since Ax:ax::ax:lr.Ax and ax:Ax::Ax:lr.ax, we have ax: lr.Ax::lr.ax:Ax. Therefore, sq.ax:sq.(lr.Ax)::sq.(lr.ax):sq.Ax::[sq.(lr.ax)-sq.ax]:[sq.Ax-sq.(lr.Ax)]. But, Ax:lr.Ax::sq.ax:sq.(lr.Ax). Hence, Ax:lr.Ax::[sq.(lr.ax)-sq.ax]:[sq.Ax-sq.(lr.Ax)].

[311] Namely, because Ax>lr.Ax, also [sq.(lr.ax)-sq.ax]>[sq.Ax-sq.(lr.Ax)].

[312] That is, if the given difference of squares is greater than the difference of the squares on the major axis and its latus rectum.

To this point, we have given [our] attention to the solution of those problems whose diorisms are immediately dependent on the diorism propositions in the Seventh book, everywhere observing the same order in which the diorisms are passed on [to us in the text]: I have not entirely persuaded myself that the diverse material had [all] been in the lost Eighth book. We do hope, however, that if we did not reach the level of Apollonius in his analyses and syntheses, we have [nevertheless] made this step, that whatever we substituted in their place[313] *would not appear inelegant to a fair reader. But even if it be that the determinate conic problems, whose analyses can be sought with little effort from these elements, are nearly endless, still, for the present, this alone was proposed to us, [namely,] that we squeeze out as much as we could of what remains of Apollonius's [work]. But if by good fortune the whole work of the author will have come to light afterwards, it will be only a minor misfortune for us, this work having [merely] been a waste of time and effort.*

313 That is, in place of Apollonius's own analyses and syntheses.

Part III
Synopsis and Appendices

Synopsis of the Contents of Halley's *Conics*, Book VIII

Although Halley's *Conics*, Book VIII is not a long work, it cannot be read in one sitting. For that reason, it seemed to me readers might find it useful to have an overview of the contents of the work so that, looking back over it, they can more easily acquire a feeling for the whole. Thus, the following synopsis.

Propositions I–IV: The Parabola and Initial Propositions for the Ellipse and Hyperbola

Proposition I requires finding the latus rectum of any diameter of a parabola, given the latus rectum of another given diameter (in particular, the axis). Proposition II requires finding a diameter of a parabola whose latus rectum equals a given line, given the latus rectum of another given diameter. These are the only propositions in the reconstruction that concern the parabola; however, they are completely consistent with the general direction of the rest of the book. For the problems in the book, as Halley conceived them, all concern relationships among diameters of conic sections, their latera recta, and, in the case of the ellipse and hyperbola, their conjugate diameters. The next two propositions, then, propositions III and IV, are the exact counterparts to proposition I for the hyperbola and ellipse, respectively.

Propositions V–XVIII: Conjugate Diameters

Aside from propositions I and II, as I have just said, all of the propositions in Halley's *Conics*, Book VIII, concern central conics, the ellipse and hyperbola. Unlike a parabola, central conics have *conjugate diameters,* pairs of diameters such that each

M.N. Fried, *Edmond Halley's Reconstruction of the Lost Book of Apollonius's Conics*, Sources and Studies in the History of Mathematics and Physical Sciences,
DOI 10.1007/978-1-4614-0146-9_8,

diameter is parallel to the ordinate direction of the other.[314] All the propositions in the reconstruction, up to the *scholium* following proposition XVIII, concern conjugate diameters, one way or another. In every proposition, the basic data are the axis of the hyperbola or ellipse and the latus rectum with respect to the axis.[315]

In propositions V (hyperbola) and VI (ellipse), accordingly, we are given the axis and its latus rectum as well as the magnitude (i.e., the length) of some other diameter; the problem, then, is (1) to find the position of the diameter and (2) to find the length, position, and latus rectum of the conjugate diameter.

Again in propositions VII (hyperbola) and VIII (ellipse), we are given the axis and its latus rectum. Here, in addition, we are given the ratio of the lengths of conjugate diameters; the task is to find the lengths of the conjugate diameters and their position in the section.

In the next four propositions, we are to find both the lengths and positions of conjugate diameters where the lengths have a given sum (propositions IX and X) or a given difference (propositions XI and XII). As always in Halley's reconstruction, the first of each pair concerns the hyperbola while the second concerns the ellipse.

Propositions XIII (hyperbola) and XIV (ellipse) ask, this time, to find the lengths and positions of a pair of conjugate diameters that contain a rectangle having a given area. For the Greek mind, it is worth remarking, this pair of problems would have represented a turn from the previous four problems. However, for one attuned to algebra, as Halley was, these six propositions form a clear nexus: conjugate diameters whose lengths have a given sum, a given difference, a given product.

The perfect symmetry, so monotonously present in the propositions until now, is lacking in the next two propositions because when it comes to the sums and differences of the squares on conjugate diameters, the subject of these propositions, the situation of the hyperbola contrasts with that of the ellipse: for the hyperbola, the difference of the squares of conjugate diameters is constant (*Conics* VII.13), while for the ellipse, it is the sum that is constant (*Conics* VII.12). Accordingly, proposition XV requires the positions and lengths of conjugate diameters of a hyperbola such that the squares of lengths of the conjugate pair have a given *sum*, while proposition XVI requires the positions and lengths of conjugate diameters of an ellipse such that the squares of lengths of the conjugate pair have a given *difference*.

As described in the introduction, the next two propositions, problems XVII (hyperbola) and XVIII (ellipse), are the first propositions that do not reflect propositions from *Conics* VII in the immediate way the others do. With the same data as in all of this first set of problems, namely, the axis and its latus rectum, the task here is to find conjugate diameters that form a given angle.

---

314 See appendix 1.

315 Sometimes this is stated in the form, "given the sides of the figure of the axis."

Scholium Special Cases of the "Application of Areas"

Ending this half of the reconstruction is a *scholium* in which Halley provides solutions to special cases of problems related to the classical theory of application of areas.[316] These lemmas, as he refers to them, treat the following problems:

(1) Given a square *S* with a given side, and a line segment *AB*, find a point *Z on the extension of AB*, such that the rectangle contained by *AZ* and *ZB* is equal to *S*;
(2) Given a square *S* with a given side, and a line segment *AB*, find a point *Z within the segment AB*, such that the rectangle contained by *AZ* and *ZB* is equal to *S*;
(3) Same as (1), except that a rectangle with given sides is given instead of a square;
(4) Same as (2), except, again, that a rectangle with given sides instead of a square is given.

In lemmas (1) and (3), the rectangle obtained, namely, that contained by *ZB* and *ABZ*, exceeds the rectangle contained by *AB* and *ZB* by a square whose side is equal to *ZB*; in lemmas (2) and (3), the rectangle obtained falls short of that contained by *ZB* and *AB* again by a square whose side is equal to *ZB*.

Propositions XIX–XXXIII: Diameters and Their Latera Recta

The second half of Halley's reconstruction contains problems that concern relationships between diameters and their latera recta. In many ways, these problems are analogous to those in the first part: diameters and latera recta must be found having a given sum, difference, ratio, containing a given area, etc. Also, as in the problems in the first half, the chief givens are the axis and its latus rectum.

The first two propositions in this section, thus, correspond exactly to proposition II. Thus, propositions XIX (hyperbola) and XX (ellipse) require finding a diameter with a latus rectum having a given length, given the axis and its latus rectum.

Propositions XXI (hyperbola) and XXII (ellipse), corresponding to propositions VII and VIII, respectively, ask for a diameter that has a given ratio with its latus rectum. And the next six propositions, propositions XXIII–XXVIII, correspond, in order, to propositions IX to XIV: the task in these is to find a diameter that has with its latus rectum a given difference (XXIII and XXIV) or sum (XXV and XXVI) or that contains a rectangle of given area (XXVII and XXVIII).

The next propositions, similarly, can be related to propositions XV and XVI in first part. Thus propositions XXIX, XXX (hyperbola), and XXXI (ellipse) ask to find a diameter and its latus rectum, the sum of whose squares is given. Propositions

[316] See appendix 1.

XXIX and XXX consider two cases for the hyperbola, one where the axis is greater than its latus rectum, and one where it is less.

The remaining two propositions XXXII (hyperbola) and XXXIII (ellipse) complement XXIX–XXXI in the obvious way, namely, they ask to find a diameter and its latus rectum, the difference of whose squares is given.

Appendix 1
Terminology and Notions from Greek Mathematics

1. The terms diameter, ordinates, figure, and latus rectum in Apollonius's Conics

Let *LAM* be a conic section. A line *AB* which bisects all lines *ML* drawn parallel to a given line is called a *diameter* of the section and the lines *ML* are called *ordinates* (in fact, Apollonius does not use a noun, but an adverbial phrase: *tetagmenōs epi tēn diametron katēxthai*, "drawn in an ordinate manner to the diameter"). If the ordinates happen to be perpendicular to the diameter, then the diameter is called the *axis* of the section. Where the diameter runs between two curves, as in the case of the "opposite sections" (what we would call the two branches of the hyperbola), it is called the *transverse diameter*.

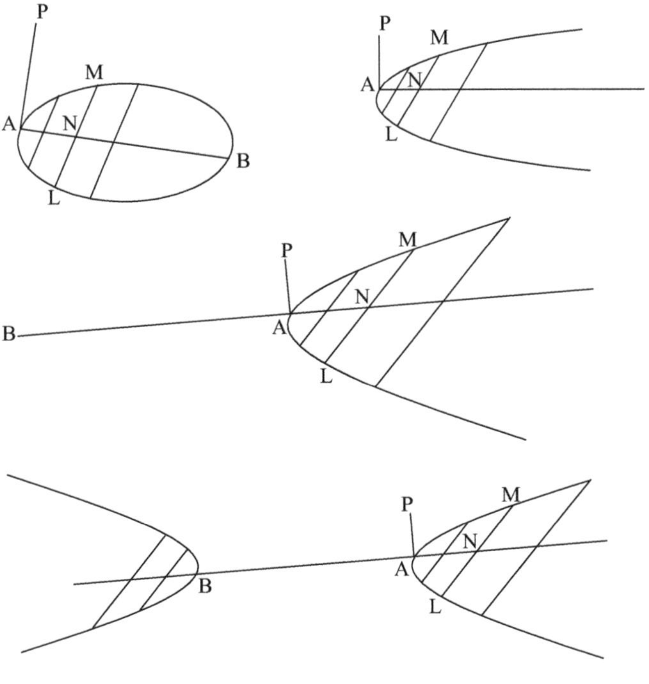

In Book I of the *Conics,* Apollonius proves that there is a line, *PA*, called the *orthia* or "upright side" (thus, *latus rectum* in Latin), such that:

i. When the section *LAM* is a *parabola,* the square on *MN* is exactly equal to the rectangle contained by *PA*, *AN*.

ii. When the section *LAM* is a *hyperbola,* the square on *MN* is equal to a rectangle *greater* than the rectangle contained by *PA*, *AN* (if the width of the rectangle is *AN*, then the excess will be a rectangle similar to the rectangle contained by *PA*, *AB*).

iii. When the section *LAM* is an *ellipse*, the square on *MN* is equal to a rectangle *less* than the rectangle contained by *PA*, *AN* (if the width of the rectangle is *AN*, then the deficiency will be a rectangle similar to the rectangle contained by *PA*, *AB*).

These properties are called the *symptōmata* of the conic sections.

The *figure* of the ellipse or hyperbola (there is no figure for a parabola) is the rectangle contained by the *orthia*, *PA*, and the *diameter*, *AB*.

2. The homologue

One of the distinctive features of *Conics*, Book VII is Apollonius's device, the *homologue*. It appears in propositions V–IX in Halley's reconstruction of Book VIII. As its name suggests, the *homologue* serves as a kind of analogue to the *orthia,* but one that is placed in the direction of the diameter instead perpendicularly to it.[317] It is constructed by dividing the diameter internally, in the case of the hyperbola, or externally, in the case of the ellipse, so that the segments obtained are in the ratio *orthia:diameter*. Apollonius generally makes two divisions in each case, one relative to each end of the diameter, and Halley, following Apollonius, does the same. In the figure below, then, *AN* and *ΓΞ* are both *homologues*:

$$AN{:}N\Gamma{::}orthia{:}diameter{::}\Gamma\Xi{:}\Xi A$$

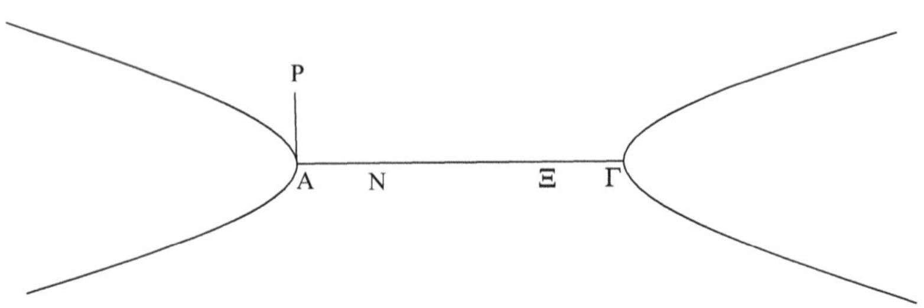

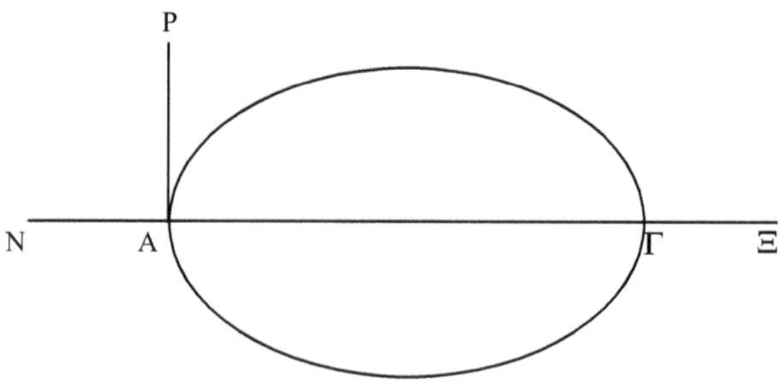

[317] See Fried (2003) for an extensive discussion.

3. Application of areas

As mentioned in the introduction, Halley's general *Scholium* treats special problems from the classical theory of application of areas. There are three basic problems in this theory, which for simplicity I will state only in terms of rectangles. The first problem, then, is this: given a line *L* and an area *A*, construct another line *W* such that the rectangle having width *W* and length *L* is equal in area to *A*.

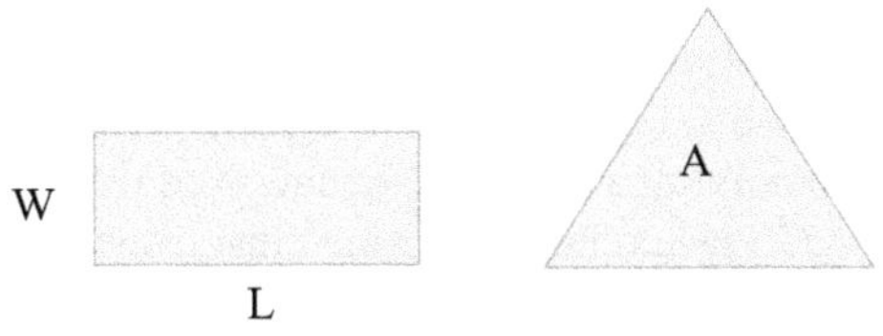

The second problem is: given a line L, an area A, and a rectangle R, construct a line W such that the rectangle having width W and length along L (though not equal to L) and equal to A exceeds the rectangle having width W and length L by a rectangle similar to R.

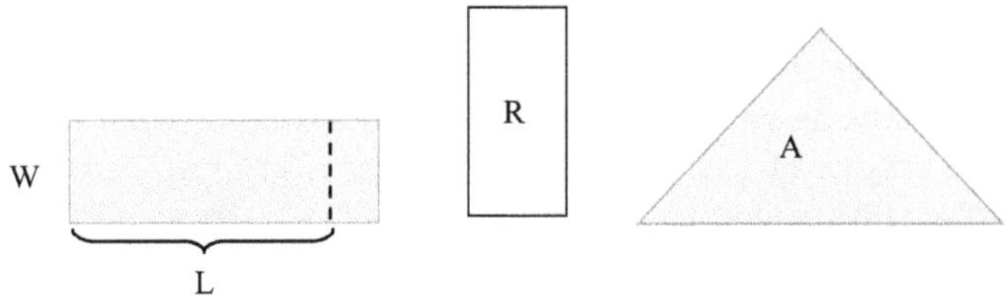

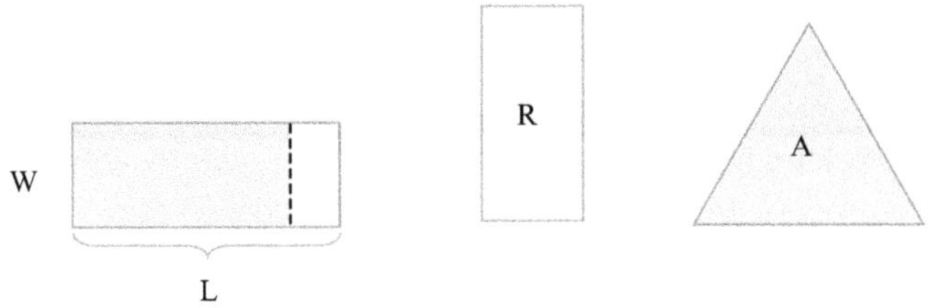

The third problem is the complement of the second: given a line L, an area A, and a rectangle R, construct a line W such that the rectangle having width W and length along L and equal to A is less than the rectangle having width W and length L by a rectangle similar to R.

To describe the construction of the rectangle (i.e., the line W) whose length lies along L and whose area equals A, the Greek mathematicians would say that they "apply an area (*parabolē tōn xōriōn*) A to a line L," thus the term "application of areas." In the first problem, the task, using the Greek terminology, was to apply an area *parabolically*, in the second, *hyperbolically*, and in the third, *elliptically*. It is from these terms that we have Apollonius's names for the conic sections, *parabola*, *hyperbola*, and *ellipse*. Indeed, comparing the definitions for the latter, one can easily see how the theory of conic sections, in the hands of Apollonius, was deeply connected to the theory of application of areas.[318]

One more word about application of areas ought to be said. It is very tempting to view application of areas in terms of algebraic equations: parabolic application becomes a matter of solving an equation of the form Lx=A; hyperbolic application becomes identified with solving a quadratic equation x(L+kx)=A (where the rectangle R is assumed to have sides in the ratio of 1:k); elliptic application, similarly, becomes identified with solving a quadratic equation of the form x(L-kx)=A. However, when considering *Greek* mathematics, one must resist that modern temptation: the Greek theory of application of areas was a geometrical theory concerned, literally, with the relationships and transformations of areas. To force that theory into an algebraic mode sweeps away not only much of the distinctiveness of Greek mathematical thought, but also of modern mathematical thought as it was taking shape in Halley's time.[319]

4. Given in magnitude and position

The phrases "given in magnitude" and "given in position" appear throughout Halley's reconstruction. They may not strike modern readers as phrases demanding much thought in themselves; however, in Greek mathematics, "to be given," one way or another, was a kind of claim important particularly (but not exclusively) in those works Pappus counts as works of analysis. The question of what it means "to be given" was the subject of Euclid's work, *Dedomena*, or in its better-known Latin name, *Data*. There we find strict definitions of "given in magnitude" and "given in position," as well as "given in ratio" and "given in form," among others. Euclid's definitions of the former are as follows:

Given in magnitude is said of figures, lines, and angles for which we can provide equals.

Given in position is said of points, lines, and angles which always hold the same place.[320]

Halley's uses of these phrases ought not be taken for granted, but seen in the light of this older Greek tradition.

---

[318] A further discussion of the idea of "application of areas," and, in particular, the connection between that theory and the theory of conic sections, can be found in Taisbak (2003a).

[319] See Klein (1968) for one of the most profound treatments of this subject.

[320] I am taking advantage of Taisbak's translations in his commentary on the *Data* (Taisbak, 2003b). In general, this is an extremely valuable book not only for understanding the idea of "being given," but also for understanding many other ideas from Greek geometry.

Appendix 2
Hippocrates's First Quadrature of a Lune

To understand Halley's remark at the end of proposition XII, one ought to review Hippocrates's first quadrature of a lune. According to Eudemus's account, as quoted by Simplicius[321] (one sees here how following sources in Greek mathematics can be messy business!), Hippocrates showed the following: suppose *ABC* is an isosceles right triangle inscribed in a semicircle, and the circular segment, *ADC*, is similar to the circular segments on the sides, *AB* and BC (which is precisely the configuration in Halley's diagram at the end of Prop. XII), then the *lune ABCD* (i.e., the curvilinear figure obtained by removing the segment *ADC* from the semicircle *ABC*) is equal in area to the triangle *ABC*.

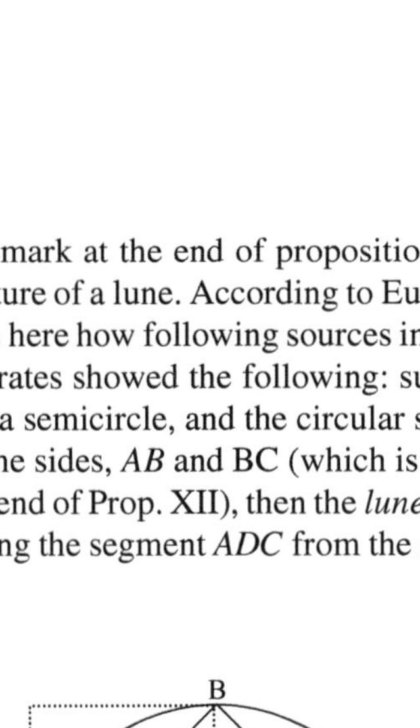

Since the circular segments are similar, they are related to one another in the same ratio as the squares on their bases. But, by the Pythagorean Theorem, the square on *AC* is equal to the sum of the squares on *AB* and *BC*. Hence, the circular segment *ADC* is equal to the sum of the circular segments on *AB* and *BC*. Therefore, since the triangle *ABC* is equal to the semicircle minus the two circular segments on *AB* and *BC*, the triangle is also equal to the semicircle minus the circular segment *ADC*, which is the *lune ABCD*. Note also, the triangle *ABC* is equal to the square on the radius, as can be seen from the dotted square added to Hippocrates's diagram.

With that, consider the diagram below adapted from Halley's diagram at the end of Prop. XII. Let the segment *EMK* be similar to $A\Lambda B$. Since sq.$E\Gamma$:sq.AZ::2:1, segment *EMK* is equal, therefore, to twice segment $A\Lambda B$, or the lozenge-shaped

[321] Simplicius, *Commentary on Aristotle's Physics* A 2 (Diels 60. 22-68).

figure *AΛBN*. Also, just as the *lune AΛBΓ* is equal to the square on *AP*, by the theorem on *lunes* above, so too the *lune EMKO* is equal to the square on *EΓ*. But the *lune EMKO* is equal to the semicircle *EOK*, or *ENK*, minus the segment *EMK*, or figure *AΛBN*. However, the semicircle *ENK* minus the figure *AΛBN* is the figure *AEKBΛ*. Therefore, as Halley observes, the figure *AEKBΛ* is equal to the square on *EΓ*.

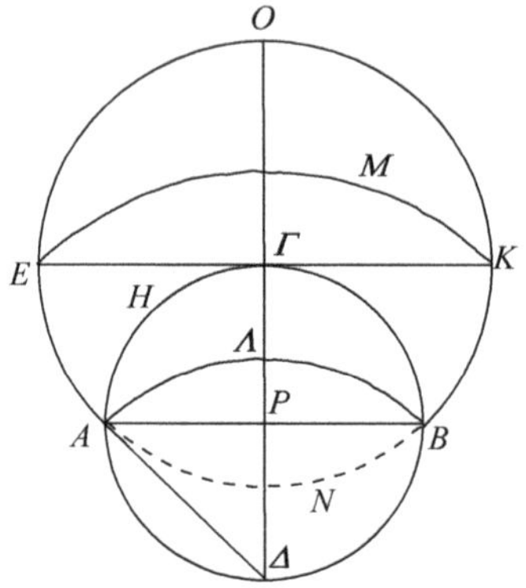

Since also the square on *EΓ* is equal to twice the square on *AP*, which is equal to the *lune AΛBΓ*, it follows that the figure *AEKBΛ* is equal to twice the *lune AΛBΓ*; therefore, also, the figure *EΓHA* is equal to the half *lune*, *AHΓΛ*.

References

1. Allen, P. (1949). Scientific Studies in the English Universities of the Seventeenth Century. *Journal of the History of Ideas, 10*(2), 219-253.
2. Allibone, T. E. (1974). Edmond Halley and the Clubs of the Royal Society. *Notes and Records of the Royal Society of London, 28*(2), 195-205.
3. Apollonius (1891, 1893). *Apollonii Pergaei quae Graece exstant cum commentariis antiquis.* I. L. Heiberg (ed.), 2 vols. Leipzig: Teubner.
4. Apollonius (1998). *Apollonius of Perga Conics: Books I–III.* Translated by R. C. Taliaferro. Dana Densmore (ed.). Santa Fe: Green Lion Press.
5. Aubrey, J. (1669–1696, 1957). *Aubrey's Brief Lives.* O. L. Dick (ed.). Ann Arbor: The University of Michigan Press.
6. Berry, A. (1961*). A Short History of Astronomy*. New York: Dover.
7. Bill, E. G. W. (1988). *Education at Christ Church Oxford, 1660–1800.* Oxford: At the Clarendon Press.
8. Boyer, C. B. (1956). *History of Analytic Geometry*. New York: Scripta Mathematica.
9. Chapman, A. (1994). Edmond Halley's Use of Historical Evidence in the Advancement of Science. *Notes and Records of the Royal Society of London, 48*(2), 167-191.
10. Chasles, M. (1860). *Les Trois Livres De Porismes D'euclide Retablis Pour La Première Fois: D'après La Notice Et Les Lemmes De Pappus.* Paris: Mallet-Bachelier.
11. Cohen, E. H.; Ross, J. S. (1985). The Commonplace Book of Edmond Halley. *Notes and Records of the Royal Society of London, 40*(1), 1-40.
12. Cook, A. (1993). Halley the Londoner. *Notes and Records of the Royal Society of London, 47*(2), 163-177.
13. Cook, A. (1998). *Edmond Halley: Charting the Heavens and the Seas*. Oxford: Clarendon Press.
14. Evans, J.; Berggren, L. (2006). *Geminos's Introduction to the Phenomena.* Princeton: Princeton University Press.
15. Fried, M. N. (2003). The Use of Analogy in Book VII of Apollonius' Conica. *Science in Context, 16*(3), 349-365.
16. Fried, M. N. (2004). A Note on the Opposite Sections and Conjugate Sections in Apollonius of Perga's *Conica. The St. John's Review, 47*(1), 91-114.
17. Fried, M. N.; Unguru, S. (2001). *Apollonius of Perga's* Conica*: Text, Context, Subtext.* Leiden, The Netherlands: Brill Academic Publishers.
18. Halley, H. (1691). A Discourse Tending to Prove at What Time and Place, Julius Caesar Made His First Descent upon Britain. *Philosophical Transactions of the Royal Society, 17*(193), 495–499.
19. Halley, H. (1694). Methodus nova accurata et facilio inveniendi radices aequationum quarumcumque generaliter sine praeria reductione. *Philosophical Transactions of the Royal Society, 18*(210), 136–148 (translation in Newton (1720)).

M.N. Fried, *Edmond Halley's Reconstruction of the Lost Book of Apollonius's Conics*, Sources and Studies in the History of Mathematics and Physical Sciences,
DOI 10.1007/978-1-4614-0146-9,

20. Halley, H. (1701). *A New and Correct Chart Shewing the Variations of the Compass in the Western and Southern Oceans as Observed in ye Year 1700 by his Maties Command by Edm. Halley*. London: Mount and Page.
21. Halley, H. (1705). *A Synopsis of the Astronomy of Comets*. London: John Senex.
22. Halley, H. (1706). *Apollonii Pergaei de Sectione Rationis Libri Duo ex Arabico MS Latine versi Accedunt ejusdem de Sectione Spatii Libri Duo restituti*. Oxford.
23. Halley, H. (1710). *Apollonii Pergaei conicorum libri octo et Sereni Antissensis de sectione cylindri et coni*. Oxford.
24. Heywood, G. (1985). Edmond Halley: Astronomer and Actuary. *Journal of the Institute of Actuaries, 112*, 279-301.
25. Heath, T. L. (1926). *The Thirteen Books of Euclid's Elements: Translated from the Text of Heiberg with introduction and Commentary*. 3 vols. Cambridge: Cambridge University Press (repr., New York: Dover Publications, Inc., 1956)
26. Hill, K. (1996). Neither Ancient nor Modern: Wallis and Barrow on the Composition of Continua. Part One: Mathematical Styles and the Composition of Continua. *Notes and Records of the Royal Society of London, 50*(2), 165-178.
27. Hill, K. (1997). Neither Ancient nor Modern: Wallis and Barrow on the Composition of Continua. Part Two: The Seventeenth-Century Context: The Struggles between Ancient and Modern. *Notes and Records of the Royal Society of London, 51*(1), 13-22.
28. Hogendijk, J. P. (1985). *Ibn al-Haytham's Completion of the Conics (Sources in the History of Mathematics and Physical Sciences 7)*. New York: Springer-Verlag.
29. Hughes, D. W. (1990). Edmond Halley: His Interest in Comets. In N. J. W. Thrower (ed.), *Standing on the Shoulders of Giants: A Longer View of Newton and Halley*, pp. 324–372. Berkeley: University of California Press.
30. Huxley, G. L. (1959). The Mathematical Work of Edmond Halley. *Scripta Mathematica, 24*, 265-273.
31. Itard, J. (1984). L'angle de contingence chez Borelli: Commentaire du livre V des Coniques d'Apollonius. In R. Rashed (ed.), *Essais D'Histoire des Mathématiques par Jean Itard*, pp.112-135. Paris: Librairie A. Blanchard.
32. Jones, A. (1986). *Pappus of Alexandria Book 7 of the Collection* (*Sources in the History of Mathematics and Physical Sciences 8*), 2 vols. New York: Springer-Verlag.
33. Klein, J. (1968). *Greek Mathematical Thought and the Origin of Algebra*. Cambridge, MA: MIT Press.
34. Lützen, J.; Purkert, W. (1994). Conflicting Tendencies in the Historiography of Mathematics: M. Cantor and H. G. Zeuthen. In E. Knobloch and D. E. Rowe (eds.), *The History of Modern Mathematics. Vol. III, Images, Ideas, Communities*, pp.1-43. Boston: Academic Press.
35. MacPike, E. F. (ed.) (1932). *Correspondence and Papers of Edmond Halley*. Oxford: At the Clarendon Press.
36. Mahoney, M. S. (1994). *The Mathematical Career of Pierre de Fermat 1601-1665*, 2nd edition. Princeton: Princeton University Press.
37. Netz, R. (1999). *The Shaping of Deduction in Greek Mathematics*. Cambridge: Cambridge University Press.
38. Newton, I. (1720). *Universal Arithmetick: Or a Treatise of Arithmetical Composition and Resolution, to which is added Dr. Halley's Method of Finding the Roots of Aequations Arithmetically*. Translation from the Latin by J. Raphson with revisions S. Cunn. London: J. Senex.
39. Newton, I. (1962). *Sir Isaac Newton's Mathematical Principles of Natural Philosophy and His System of the World*, 2 volumes. F. Cajori (ed.). Berkeley: University of California Press.
40. Ronan, C. (1969). *Edmond Halley: Genius in Eclipse*, London: Macdonald.
41. Suttle, E. F. A. (1940). Henry Aldrich, Dean of Christ Church. *Oxoniensia, 5,* 115-139.
42. Taisbak, C. M. (2003a). Exceeding and Falling Short: Elliptical and Hyperbolic Application of Areas. *Science in Context,16*(3), 299-318.
43. Taisbak, C. M. (2003b). *Dedomena: Euclid's Data, or, The Importance of Being Given*. Copenhagen: Museum Tusculanum Press.
44. Toomer, G. (1970). Apollonius of Perga. In C. C. Gillespie (ed.), *Dictionary of Scientific Biography*, I, pp. 179-193. New York: Charles Scribner and Sons.

45. Toomer, G. J. (1990). *Apollonius Conics Books V to VII*, 2 vols. New York: Springer-Verlag.
46. Traub, J.F. (1964). *Iterative Methods for the Solution of Equations*. Englewood Cliffs: Prentice Hall.
47. Turnbull, H. W. (1945). *The Mathematical Discoveries of Newton*. London: Blackie.
48. Unguru, S. (1976). A Very Early Acquaintance with Apollonius of Perga's Treatise on Conic Sections in the Latin West. *Centaurus*, *20*, 112-128.
49. Unguru, S. (1979). History of Ancient Mathematics: Some Reflections of the State of the Art. *Isis*, *70*, 555–565.
50. Unguru, S. (1974). Pappus in the Thirteenth Century in the Latin West. *Archive for History of Exact Sciences, 13*(4), 307-324.
51. Warntz, W.; Wolff, P. (1971). *Breakthroughs in Geography*. New York: New American Library.
52. Zeuthen, H. G. (1886). *Die Lehre von den Kegelschnitten im Altertum*. Kopenhagen (repr., Hildesheim: Georg Olms Verlagsbuchhandlung, 1966).

Index

M.N. Fried, *Edmond Halley's Reconstruction of the Lost Book of Apollonius's Conics*, Sources and Studies in the History of Mathematics and Physical Sciences, DOI 10.1007/978-1-4614-0146-9,

GPSR Compliance
The European Union's (EU) General Product Safety Regulation (GPSR) is a set of rules that requires consumer products to be safe and our obligations to ensure this.

If you have any concerns about our products, you can contact us on

ProductSafety@springernature.com

In case Publisher is established outside the EU, the EU authorized representative is:

Springer Nature Customer Service Center GmbH
Europaplatz 3
69115 Heidelberg, Germany

www.ingramcontent.com/pod-product-compliance
Ingram Content Group UK Ltd.
Pitfield, Milton Keynes, MK11 3LW, UK
UKHW052315141026
476617UK00003B/37
* 9 7 8 1 4 6 1 4 0 1 4 5 2 *